AF779117

Recent Advances on Safe Maritime Operations under Extreme Conditions

Recent Advances on Safe Maritime Operations under Extreme Conditions

Editors

Pentti Kujala
Spyros Hirdaris
Martin Bergström

MDPI • Basel • Beijing • Wuhan • Barcelona • Belgrade • Manchester • Tokyo • Cluj • Tianjin

Editors
Pentti Kujala
Marine and Arctic Technology
Aalto University
Espoo
Finland

Spyros Hirdaris
Marine and Arctic Technology
Aalto University
Espoo
Finland

Martin Bergström
Marine and Arctic Technology
Aalto University
Espoo
Finland

Editorial Office
MDPI
St. Alban-Anlage 66
4052 Basel, Switzerland

This is a reprint of articles from the Special Issue published online in the open access journal *Applied Sciences* (ISSN 2076-3417) (available at: www.mdpi.com/journal/applsci/special_issues/safe_maritime_operations).

For citation purposes, cite each article independently as indicated on the article page online and as indicated below:

LastName, A.A.; LastName, B.B.; LastName, C.C. Article Title. *Journal Name* **Year**, *Volume Number*, Page Range.

ISBN 978-3-0365-1919-7 (Hbk)
ISBN 978-3-0365-1918-0 (PDF)

© 2021 by the authors. Articles in this book are Open Access and distributed under the Creative Commons Attribution (CC BY) license, which allows users to download, copy and build upon published articles, as long as the author and publisher are properly credited, which ensures maximum dissemination and a wider impact of our publications.
The book as a whole is distributed by MDPI under the terms and conditions of the Creative Commons license CC BY-NC-ND.

Contents

About the Editors vii

Preface to "Recent Advances on Safe Maritime Operations under Extreme Conditions" ix

Pentti Kujala, Spyros Hirdaris and Martin Bergström
Special Issue on the Recent Advances in Safe Maritime Operations under Extreme Conditions
Reprinted from: *Applied Sciences* **2021**, *11*, 5789, doi:10.3390/app11135789 1

Bernt J. Leira, Wei Chai and Gowtham Radhakrishnan
On Characteristics of Ice Ridges and Icebergs for Design of Ship Hulls in Polar Regions Based on Environmental Design Contours
Reprinted from: *Applied Sciences* **2021**, *11*, 5749, doi:10.3390/app11125749 5

Martin Bergström and Pentti Kujala
Simulation-Based Assessment of the Operational Performance of the Finnish–Swedish Winter Navigation System
Reprinted from: *Applied Sciences* **2020**, *10*, 6747, doi:10.3390/app10196747 29

Thomas Browne, Rocky Taylor, Brian Veitch, Pentti Kujala, Faisal Khan and Doug Smith
A Framework for Integrating Life-Safety and Environmental Consequences into Conventional Arctic Shipping Risk Models
Reprinted from: *Applied Sciences* **2020**, *10*, 2937, doi:10.3390/app10082937 57

Wei Li, Xiao Liang, Jianguo Lin, Ping Guo, Qiang Ma, Zhenpeng Dong, Jiamin Liu, Zhenhe Song and Hengqi Wang
Numerical Simulation of Ship Oil Spill in Arctic Icy Waters
Reprinted from: *Applied Sciences* **2020**, *10*, 1394, doi:10.3390/app10041394 81

Meng Zhang, Karl Garme, Magnus Burman and Li Zhou
A Numerical Ice Load Prediction Model Based on Ice-Hull Collision Mechanism
Reprinted from: *Applied Sciences* **2020**, *10*, 692, doi:10.3390/app10020692 101

Sabina Idrissova, Martin Bergström, Spyros E. Hirdaris and Pentti Kujala
Analysis of a Collision-Energy-Based Method for the Prediction of Ice Loading on Ships
Reprinted from: *Applied Sciences* **2019**, *9*, 4546, doi:10.3390/app9214546 123

About the Editors

Pentti Kujala

Since 2006, Pentti Kujala has been a Professor of Marine Technology (safety) at the Aalto University, School of Engineering. Since 2012, he has been the Head of the Marine Technology research group, and from May 2017, he has also been the Vice Dean of Research for the School of Engineering. He chairs a Center of Excellence for Arctic Shipping and Operations (CEPOLAR) funded by Lloyd's Register Foundation (2013–2021). His main research interests have been devoted to the safety and risk analysis of marine operations both in open water and in ice, and the development of innovative structural solutions for various types of ships. In total, he has published over 300 papers with approximately 3900 citations and has a Scopus h-index of 34.

Spyros Hirdaris

Spyros Hirdaris is an Associate Professor of Maritime Technology. In his research, he combines knowledge from advanced ship and safety science, marine hydrodynamics and structures for the prediction of sea loads, safety, and the performance of ships and offshore structures operating in extreme conditions. He completed his PhD in 2002 on Ship Science (Hydroelasticity of Ships) at the University of Southampton. From 2004 to 2018, he worked internationally for Lloyd's Register. This work involved research and product development, planning and strategy for R&D, consultancy, and new marine construction activities. He is a Chartered Engineer, Fellow of the Royal Institution of Naval Architects (UK), and a Member of the Society of Naval Architects and Marine Engineers (USA). He has served in the International Ship and Offshore Structures Congress as a member and chairman of various committees with a focus on sea loads and responses since 2008.

Martin Bergström

Martin Bergström, Ph.D., M.Sc (Tech.), is a Naval Architect and Postdoctoral Researcher at Aalto University in Finland. He completed his Ph.D. at NTNU, the Norwegian University of Science and Technology, in 2017. In his research, he develops tools and approaches towards safe and efficient winter navigation. Currently, he primarily focuses on the simulation and analysis of winter navigation systems to study the implications of climate change and Efficiency Design Index (EEDI) regulations, among others. In addition, he contributes to the development of methods supporting the goal-based design of ice-class ships following the IMO Polar Code.

Preface to "Recent Advances on Safe Maritime Operations under Extreme Conditions"

Increased activity in the Arctic involves hazards such as a harsh environment, especially the ice cover and cold temperature, remoteness, a lack of infrastructure, and lack of information about bathymetry, among other challenges. Ice cover is also highly variable and dynamic, with increasing variation in the future; due to the changing effects of the world climate, the ice conditions on all ice-covered areas are also under dynamic change. The effects on Arctic operations are complicated issues to address. The remoteness of Arctic areas means that in the case of an accident, the search and rescue (SAR) capabilities are low. Fairways are not marked very extensively; therefore, soundings taken for charting are relatively scarce. These polar area hazards are compounded by the fact that the rate of recovery of nature is slow, meaning that environmental hazards are exacerbated.

This book presents a collection of papers which focus on scenario-based risk management for polar shipping and risk-based guidelines. Topics discussed by internationally renowned authors involve the description of ice conditions, ship–ice contact and associated loading, numerical models/idealizations, definitions of limit states for ice-strengthened structures, and holistic risk/reward analyses for polar navigation. From an overall perspective, this work contributes toward international research efforts to safeguard maritime operations in ice-infested waters. This book may be of interest for maritime industry practitioners, regulators, academics, researchers and postgraduate students.

Part of the work presented in this book has been sponsored by Lloyd's Register Foundation (LRF). Their financial support is acknowledged with gratitude. LRF helps to protect life and property by supporting engineering-related education, public engagement, and the applications of research.

Pentti Kujala, Spyros Hirdaris, Martin Bergström
Editors

applied sciences

MDPI

Editorial

Special Issue on the Recent Advances in Safe Maritime Operations under Extreme Conditions

Pentti Kujala *, Spyros Hirdaris and Martin Bergström

Department of Mechanical Engineering, School of Engineering, Aalto University, 00076 Aalto, Finland; spyros.hirdaris@aalto.fi (S.H.); martin.bergstrom@aalto.fi (M.B.)
* Correspondence: pentti.kujala@aalto.fi

1. Introduction

The increased activity in the Arctic involves hazards such as a harsh environment, especially the ice cover and cold temperature, remoteness and lack of infrastructure, and lack of information about bathymetry, among others. Ice cover is also highly variable and dynamic, with increasing variation in the future as, due to the changing effects of the world climate, the ice conditions on all ice-covered areas are also under dynamic change. This effect on Arctic operations is a complicated problem to solve. The remoteness of the Arctic areas means that in case of an accident, the search and rescue (SAR) capability is low. Further, the fairways are not marked very extensively, and the soundings taken for charting are relatively scarce. These polar area hazards are compounded by the fact that the rate of recovery of nature is slow, meaning that environmental hazards are made more serious.

The focus of this Special Issue is research related to the recommended practices of scenario-based risk management for polar shipping and risk-based guidelines considering holistically the impact of risks on ice-infested waters. These include: (a) definition of ice conditions, (b) ship–ice contact, (c) numerical models/idealizations, (d) definition of limit states for ice-strengthened structures, and (e) holistic risk–reward analysis for polar navigation.

check for updates

Citation: Kujala, P.; Hirdaris, S.; Bergström, M. Special Issue on the Recent Advances in Safe Maritime Operations under Extreme Conditions. *Appl. Sci.* **2021**, *11*, 5789. https://doi.org/10.3390/app11135789

Received: 24 May 2021
Accepted: 14 June 2021
Published: 22 June 2021

Publisher's Note: MDPI stays neutral with regard to jurisdictional claims in published maps and institutional affiliations.

Copyright: © 2021 by the authors. Licensee MDPI, Basel, Switzerland. This article is an open access article distributed under the terms and conditions of the Creative Commons Attribution (CC BY) license (https://creativecommons.org/licenses/by/4.0/).

2. Summary of Contents

This special includes six original research articles on recent advances in safe maritime operations under extreme conditions.

Bergström and Kujala [1] present a discrete event simulation-based approach for assessing the operating performance of the Finnish–Swedish winter navigation system (FSWNS) under different operating scenarios. Different operating scenarios are specified in terms of ice conditions, the volume of maritime traffic, number of icebreakers (IBs), and regulations such as the energy efficiency design index (EEDI). Case studies indicate that, unless the number of IBs is increased, the EEDI regulations may result in a significant increase in both the number of instances of IB assistance and the cumulated IB waiting times.

Brown et al. [2] present a new framework that augments the current POLARIS methodology to model consequences. It has been developed on the premise that vessels of a given class with higher potential life-safety, environmental, or socio-economic consequences should be operated more conservatively. The framework supports voyage planning and real-time operational decision making through assignment of operational criteria based on the likelihood of ice-induced damage and the potential consequences. The objective of this framework is to enhance the safety of passengers and crews and the protection of the Arctic environment and its stakeholders.

Li et al. [3] present a three-dimensional numerical simulation model of an oil spill for application in emergency treatment methods under icy water conditions. The combined effects of wind, wave, current, and ice implemented in their model correspond to Arctic Ocean conditions. A discrete element method combined with an overset grid was adopted to track the trajectory movements of an oil film with medium-density ice floes and simulate the flow field of moving ice of large displacement in six degrees of freedom (6DOF). The probability of oil spill area extensions were estimated by a response surface method (RSM). The results show a reduced risk of pollution in icy water conditions and greater drift action of oil film. Accordingly, the spraying location and quantity of oil dispersant could be rapidly specified.

Zhang et al. [4] present a simplified numerical model to predict ice impact forces acting on a ship's hull in level ice conditions. The model is based on ice–hull collision mechanisms and the essential ice-breaking characteristics. The two critical ice failure modes, localized crushing and bending breaking, are addressed. An energy method is used to estimate the crushing force and the indentation displacement for different geometry schemes of ice–ship interaction. The ice bending breaking scenario is taken as a semi-infinite plate under a distributed load resting on an elastic foundation. An integrated complete ice–hull impact event is introduced with ice failure modes and breaking patterns. Impact location randomness and number of broken ice wedges are considered in order to establish a stochastic model. This model can be used to predict ice impact loads, and to establish a link between design parameters (ice properties and ship geometry) and structural loads.

Idrissova et al. [5] analyze the effect of the assumptions behind the Popov method by comparing ice load predictions, calculated by the method, with corresponding full-scale ice load measurements. The findings indicate that assumptions concerning the modelling of the ship–ice collision scenario, the ship–ice contact geometry, and the ice conditions among others, significantly affect how well the ice load prediction agrees with the measurements.

Leira et al. [6] illustrate the most likely combinations of the statistical properties of the parameters that characterize ice ridges and icebergs, using the so-called environmental contour method. Probabilistic models of the key parameters that govern the ship–ice interaction process are introduced. Subsequently, the procedure referred to as the inverse reliability method (IFORM) is applied for identification of the environmental contour. Different forms (i.e., dimensions) of environmental contours are generated to reflect the characteristics of the interaction process. Furthermore, the effect of an increasing correlation between the basic parameters is studied. In addition, the increase in the design parameter values for increasing encounter frequencies is illustrated.

3. Perspectives on Future Research

Climate change is expected to cause rapid changes in sea ice conditions, increasing the volume of shipping in icy waters. The current experience-based determination of ice-induced loads on ships, primarily derived from historical data, are not suitable for handling the related challenges. Therefore, new approaches are urgently needed, based on detailed full-scale measurements, as well as an in-depth understanding of the physics of the ship–ice interaction. The main challenges are related to modelling the icebreaking process, due to its stochastic nature, and due to large variations in the prevailing ice conditions and material properties. Despite significant research efforts, no clear link between ice-induced loads and the prevailing ice conditions exists. The resulting uncertainty in ship design translates into significant risks to people and the environment.

Author Contributions: Writing—original draft preparation; P.K., writing—review and editing; S.H. and M.B. All authors have read and agreed to the published version of the manuscript.

Funding: This research received no external funding.

Conflicts of Interest: The authors declare no conflict of interest.

References

1. Bergström, M.; Kujala, P. Simulation-Based Assessment of the Operational Performance of the Finnish–Swedish Winter Navigation System. *Appl. Sci.* **2020**, *10*, 6747. [CrossRef]
2. Browne, T.; Taylor, R.; Veitch, B.; Kujala, P.; Khan, F.; Smith, D. Framework for Integrating Life-Safety and Environmental Consequences into Conventional Arctic Shipping Risk Models. *Appl. Sci.* **2020**, *10*, 2937. [CrossRef]

3. Li, W.; Liang, X.; Lin, J.; Guo, P.; Ma, Q.; Dong, Z.; Liu, J.; Song, Z.; Wang, H. Numerical Simulation of Ship Oil Spill in Arctic Icy Waters. *Appl. Sci.* **2020**, *10*, 1394. [CrossRef]
4. Zhang, M.; Garme, K.; Burman, M.; Zhou, L. A Numerical Ice Load Prediction Model Based on Ice-Hull Collision Mechanism. *Appl. Sci.* **2020**, *10*, 692. [CrossRef]
5. Idrissova, S.; Bergström, M.; Hirdaris, E.S.; Kujala, P. Analysis of a Collision-Energy-Based Method for the Prediction of Ice Loading on Ships. *Appl. Sci.* **2019**, *9*, 4546. [CrossRef]
6. Leira, B.J.; Chai, W.; Radhakrishnan, G. On Characteristics of Ice Ridges and Icebergs for Design of Ship Hulls in Polar Regions Based on Environmental Design Contours. *Appl. Sci.* **2021**, *11*, 5749. [CrossRef]

applied sciences

MDPI

Article

On Characteristics of Ice Ridges and Icebergs for Design of Ship Hulls in Polar Regions Based on Environmental Design Contours

Bernt J. Leira [1,*], Wei Chai [2,*] and Gowtham Radhakrishnan [1]

[1] Department of Marine Technology, Norwegian University of Science and Technology, 7052 Trondheim, Norway; gowtham.radhakrishnan@ntnu.no
[2] School of Transportation, Wuhan University of Technology, Wuhan 400063, China
* Correspondence: bernt.leira@ntnu.no (B.J.L.); chaiwei@whut.edu.cn (W.C.); Tel.: +47-73595989 (B.J.L.); +86-18807151734 (W.C.)

Abstract: Ice ridges and icebergs generally pose a major threat to both ships and offshore facilities that operate in Polar regions. In many cases these features will govern the structural design loads associated with the Ultimate Limit State (ULS) and the Accidental Limit State (ALS). In general, a large number of load cases must be considered in order to ascertain an adequate structural resistance. Alternatively, conservatively high values of the relevant design parameters can be applied, which implies cost penalties. Accordingly, it is natural to consider methods that can serve to reduce the number of relevant load cases. Based on relevant information about the statistical properties of the parameters that characterize ice ridges and icebergs, the most likely combinations of these parameters for design purposes are highly relevant. On this background, the so-called environmental contour method is applied. Probabilistic models of the key parameters that govern the ship and ice interaction process are introduced. Subsequently, the procedure referred to as inverse reliability methods (IFORM) is applied for identification of the environmental contour. Different forms (i.e., dimensions) of environmental contours are generated to reflect the characteristics of the interaction process. Furthermore, the effect of an increasing correlation between the basic parameters is studied. In addition, the increase of the design parameter values for increasing encounter frequencies is illustrated.

Keywords: design contours; ice ridge; iceberg; Polar ships; environmental conditions

check for updates

Citation: Leira, B.J.; Chai, W.; Radhakrishnan, G. On Characteristics of Ice Ridges and Icebergs for Design of Ship Hulls in Polar Regions Based on Environmental Design Contours. *Appl. Sci.* **2021**, *11*, 5749. https://doi.org/10.3390/app11125749

Academic Editors: Pentti Kujala, Spyros Hirdaris and Martin Bergström

Received: 16 May 2021
Accepted: 15 June 2021
Published: 21 June 2021

Publisher's Note: MDPI stays neutral with regard to jurisdictional claims in published maps and institutional affiliations.

Copyright: © 2021 by the authors. Licensee MDPI, Basel, Switzerland. This article is an open access article distributed under the terms and conditions of the Creative Commons Attribution (CC BY) license (https://creativecommons.org/licenses/by/4.0/).

1. Introduction

The recent decades have seen an increasing interest and resulting demand for development of Arctic ships and offshore structures. These are used for exploration and extraction of natural resources, and for navigation throughout the Arctic and Antarctic corridors [1,2]. A number of different ice features will generally be of concern in these areas, such as ice ridges and icebergs. These features are assumed to represent a major hazard to the integrity of ship hulls and may accordingly govern the design loads [3].

Design of ships for Polar regions is mainly based on rules and regulations, such as the International Maritime Organization (IMO) Polar code, the Finnish–Swedish Ice Class Rules (FSICR), the International Association of Classification Societies (IACS) Polar Class rules, etc. These codes and rules are primarily based on experience and deterministic solutions, and they are attractive in connection with practical design due to their simplicity [4]. However, ship-ice loads are random by nature [5,6]. This randomness is due to the inherent variability of ice conditions (such as the physical ice characteristics including the mechanical properties) and also by the great span of ship-ice interaction processes [7]. Accordingly, probabilistic methods are required in order to account for the inherent randomness of the ship-ice loads. Hence, reliability-based design methods based on proper representation of the statistical variation of ice loads could serve to supplement current rule-based design methods.

Regarding structural design principles, criteria corresponding to the Ultimate Limit State (ULS) are intended to ensure that no extensive structural damage is likely to take place during the intended lifetime of the structure. For Polar ships, criteria associated with the ULS condition imply that the ship should have the capacity to resist the ice load actions corresponding to a specific return period (or a specific exceedance probability) without critical damage taking place. This applies both to the local and the global load effects within the vessels [8]. In general, it is assumed that the most accurate approach for estimation of the extreme ice loads is to perform a full long-term response analysis. As part of this analysis, each individual ice condition is weighted by the associated probability of occurrence. However, this type of analysis is usually time-consuming since high-fidelity numerical response analyses are usually involved [9].

Regarding the Accidental Limit State, a range of different design scenarios also need to be analyzed. Since highly nonlinear structural response behavior is generally involved, the associated computational efforts easily become tremendous. Particularly for the case of so-called shared energy design, complex dynamic interaction processes involving both the ice mass and the floating structure must be modeled in a proper way. This requires careful building of computational models and numerical algorithms [10].

In order to make ULS and ALS design procedures more efficient, the environmental contour method offers an effective alternative to the full long-term analysis. The contour will then represent a set of environmental conditions (presently corresponding to ice ridge and iceberg characteristics) for a given vessel lifetime [11]. The extreme response which corresponds to the ice loads for this specified "return period" can subsequently be calculated by identifying the critical ice conditions along the generated environmental contour. The advantage of this approach lies in the fact that the environmental characteristics are uncoupled from the structural response during the first step of the analysis [12]. Numerical simulations (in some cases supplemented by experiments) are then only required for a very limited number of conditions that are located on the contour in order to identify the most critical structural response level.

The environmental contour is generally identified by application of the IFORM (inverse first order reliability method). The calculations are based on the joint probability distribution associated with the relevant environmental parameters [13]. The concept of environmental contours has been widely used in the field of offshore engineering, e.g., in connection with hydrodynamic loading [11]. Typically, the joint statistics of wave height and peak period are described by a conditional modeling approach. For a given return period, a circle with the desired radius is created in a normalized Gaussian space, and based on an inverse mapping (IFORM) involving, e.g., the Rosenblatt transformation, the circle is mapped into the corresponding environmental contour as a continuous functional relationship between the relevant physical parameters. In addition to the environmental contours based on IFORM, alternative similar approaches are also available [14,15].

In this work, application of environmental contours is considered in relation to ice ridge and iceberg characteristics. Estimation of the extreme response due to the associated ice loads is also addressed. As a first step, key parameters associated with ice ridges and icebergs for estimation of the resulting ship-ice loads are identified by consideration of the ship–ice interaction process. Statistical models are introduced in order to characterize the variability of these parameters. By application of the inverse cumulative distribution functions of these key parameters, environmental design contours for a given return period are established.

The main objective of the present paper is to extend the methodology associated with recent application of environmental contour line methods such that they can also manage to include accidental load conditions. It is aimed at establishing a unified basis for how this can be accomplished.

In order to achieve this, relevant structural design and acceptance criteria must also be revisited since these will be different for accidental conditions as compared to ice loads with a more "regular" rate of occurrence (i.e., shorter return periods). Accidental scenarios can be widely different, and relevant design criteria are accordingly not well-established. Recent design codes also tend to move towards a more goal-based approach rather than the prescriptive rules of today, which implies less specific and detailed design requirements.

It is believed that the contour design approach will lead to increasingly efficient identification of critical design conditions. This in turn also implies great savings in terms of number of complex and time-consuming nonlinear numerical response-analyses that will be required.

2. The Ultimate and the Accidental Limit States

As discussed above, the "permissible" structural damage is very different for the Ultimate versus the Accidental Limit State. This is thoroughly discussed, e.g., in [4,7,10]. A further illustration is provided by Figure 1, where relative ULS deformations are exemplified to the left and ALS deformations to the right. It is seen that the deformation of the ice is strongly influenced by the structural strength and the corresponding structural behavior.

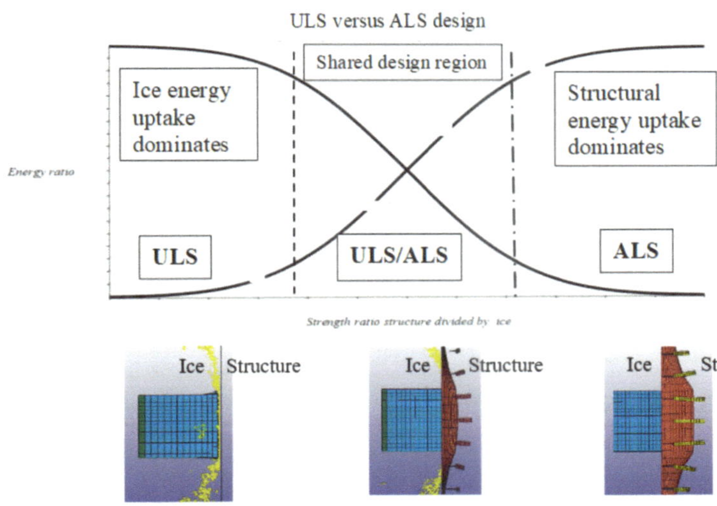

Figure 1. ULS, shared design, and ALS. The vertical axis shows energy absorbed by structure divided by energy absorbed by ice feature (energy ratio).

The relative uptake of energy by the ice feature versus the structure is shown in the lower part of the figure—for the ULS design the impact energy is mainly absorbed by the ice feature, while for the ALS design the structure typically absorbs most of the total kinetic energy. This total kinetic energy for the ship–ice system just prior to the impact event can be expressed as follows [10,16]:

$$E_S = \frac{1}{2}(m_s + a_s)v_s^2 \frac{\left(1 - \frac{v_i}{v_s}\right)^2}{1 + \left(\frac{m_s + a_s}{m_i + a_i}\right)} \quad (1)$$

where m_s and a_s are the dry mass and the added mass of the ship (e.g., [17]), respectively; m_i and a_i denote the corresponding quantities for the ice feature; v_s is the corresponding speed of the ship, and v_i that of the ice feature.

The energy balance (by disregard remaining energy in the system after impact, e.g., due to the ship or the ice feature having nonzero velocities) is then expressed as the kinetic energy before the impact being equal to the sum of the energy absorbed by the structure and that absorbed by the ice feature:

$$E_s = E_{s,ice} + E_{s,structure} \tag{2}$$

where the first term represents the energy absorbed by the ice feature and the second term corresponds to the energy that is absorbed by the structure.

The absorbed energies correspond to the areas under the respective force–displacement curves. The energy uptake by the structure versus the ice feature is further illustrated in Figure 2 [18] where the results correspond to a shared energy scenario. In Figure 2a, the amount of energy absorbed in the case of an ice floe impact is shown with the ice contribution to the left and the structural contribution to the right (shaded areas). In Figure 2b, load–displacement curves corresponding to increasing mass of the ice feature are shown where the blue curve corresponds to the smallest mass (288 tons) and the violet curve corresponds to the largest mass (1500 tons). It is seen that the maximum deformation increases for both the structure and for the ice feature as the impacting mass increases. This implies a corresponding increase of the sum of the absorbed energies that are required in order to balance the initial kinetic energy. In addition to initial kinetic energy, it is found that the local shape of the ice feature for the part where the impact occurs also has a strong influence on the resulting contact force and relative amount of energy absorption.

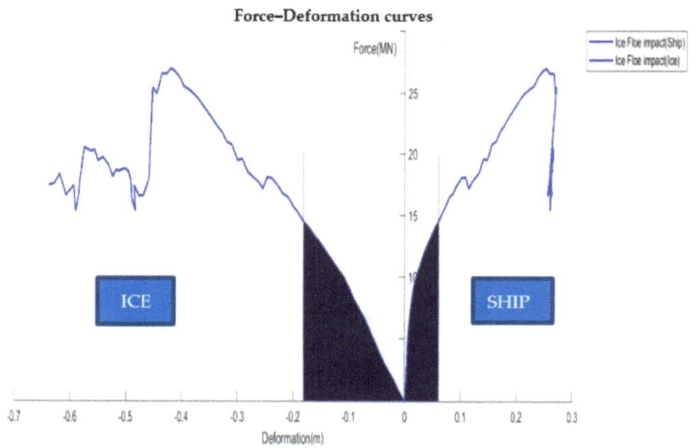

(a)

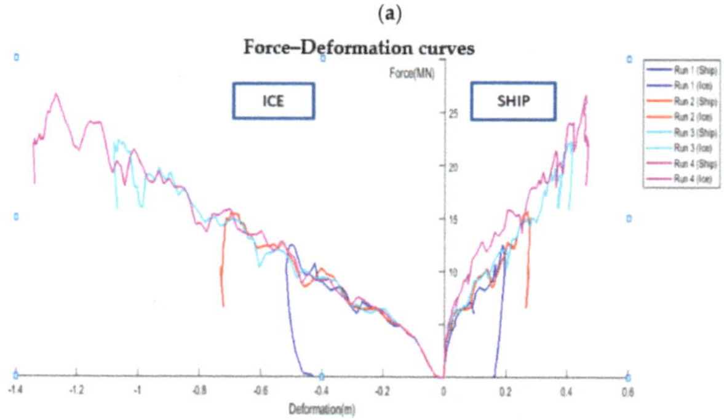

(b)

Figure 2. Energy absorption and examples of force–displacement curves: (**a**) impact by ice floe and (**b**) comparison of load–displacement curves and energy absorption for different impact scenarios (blue curves: ice mass is 288 tons; red curves: ice mass is 500 tons; magenta curves: ice mass is 1000 tons; violet curves: ice mass is 1500 tons).

3. Environmental Contour Method

The main idea behind design contours is to identify the most relevant environmental conditions corresponding to a specific return period (or equivalently specified in terms of a probability of exceedance). Subsequently, response analyses are performed for a subset of these conditions. This saves computation time as compared to a more systematic and more accurate long-term response analysis for which the whole range of possible environmental conditions must be considered.

In order to perform such a full long-term analysis, the following two-step procedure is relevant:

I. The ULS or ALS criterion is first expressed on the following form:

$$G(y_c, \mathbf{S}) = y_c - Y(\mathbf{S}) \quad (3)$$

Here, $G(\cdot)$ designates the mechanical failure function; the n-dimensional vector $\mathbf{S} = (S_1, S_2, \ldots, S_n)^{\mathrm{T}}$ denotes the basic design variables for which the joint probability density function (PDF) is assumed to be known, which is denoted by $f_S(s)$. $Y(\mathbf{S})$ is the internal load effect in the structure. The quantity y_c is the corresponding design capacity (which can be a function of several deterministic parameters and/or additional random variables related to the structural properties). Typically, the load effect and structural capacity is formulated in terms of stresses, forces, and bending moments.

As an example in relation to the ULS, y_c may, e.g., correspond to the design value of the yield stress of the steel for the ship hull and $Y(S)$ will then represent the maximum stress load effect in the hull (as a function of the values of the basic design variables). However, since this corresponds to a very strict design criterion, it is more relevant instead to take y_c to designate a maximum permissible plastic strain limit (which for a particular hull member can be translated into a maximum plastic deformation limit). The quantity $Y(\mathbf{S})$ then designates the maximum strain load effect in the hull. Similarly for the ALS, y_c may refer to a critical plastic strain limit which generally is higher than for the ULS. Depending on the accidental scenario, it could even correspond to the fracture strain of the hull material or a certain maximum allowable fracture damage.

However, in the following, a formulation is also provided where both the load effect and capacity are defined in terms of energy levels, i.e., by applying the absorbed impact energy as the load effect and the critical impact energy (corresponding to a given failure mode) as the capacity. This is in accordance with the impact analysis approach outlined in the previous section.

II. The failure probability p_f of the structure corresponding to a given time in operation can next be calculated as:

$$p_f(y_c) = \int_{G(y_c,\mathbf{S}) \leq 0} f_S(s)ds \quad (4)$$

Basically, this integral expresses the probability content that corresponds to the "volume" of the failure domain in the space which is spanned by the relevant random variables. The boundary of the failure domain is defined as the surface that is obtained by setting the mechanical limit state function in Equation (3) equal to zero.

The integral in Equation (4) can be only be expressed in closed form for cases where the joint PDF and the limit state function $G(y_c, \mathbf{S})$ are given. However, this is quite rare in practice since the load effect $Y(\mathbf{S})$ for a given value of the environmental parameter vector $\mathbf{S}$ usually needs to be obtained by means of numerical simulation and/or by experiments.

Evaluated of the integral can also be made by application of the Monte Carlo simulation technique (MCS) or by other reliability methods such as the first order or second order reliability methods (FORM/SORM) [19]. These approaches are based on the joint PDF and the failure surface $G(y_c, \mathbf{S})$, which are combined in order to establish a transformation into a normalized space of independent, standard Gaussian variables (which is frequently referred to as the U-space). In the FORM approach, the failure surface in normalized space is represented by the tangent plane at the so-called design point. The corresponding probability of failure is approximated as:

$$p_f(y_c) \approx \Phi(-\beta) \tag{5}$$

where Φ represents the standard normal cumulative distribution function (CDF); β denotes the reliability index, and this also corresponds to the distance between the design point to the origin in normalized space.

It is emphasized that the relationship between the failure probability and the standard cumulative Gaussian distribution function is due to the transformation into normalized and independent variables. Furthermore, this is a first order approximation based on linearization of the mechanical limit state function. Second order corrections to this approximation can be obtained, e.g., by the so-called SORM approximation. Generally, various types of Monte Carlo simulation techniques are also applied as a supplement to (or sometimes instead of) the approximation in Equation (5).

However, determination of the design load effect, y_N, which corresponds to the capacity that is required in order to withstand the loads associated with an N-year return period generally requires iterative reliability calculations for different values of y_c.

Accordingly, it is highly relevant to simplify Step II of the long-term analysis procedure, which is precisely the objective of environmental contour methods. The design contour for the N-year return period is frequently established by means of the inverse FORM (i.e., IFORM) approach [13,20]. The probability of failure, $p_f(y_N)$ is then first specified in order to generate the environmental contour that corresponds to that probability. An n-dimensional sphere with the radius β_F, which is obtained by means of the specified failure probability as expressed by Equation (6), is then first created in the normalized Gaussian space:

$$\beta_F = \Phi^{-1}(1 - p_f(y_N)) \tag{6}$$

This sphere is subsequently transformed into the physical parameter space to yield the design contour. The mapping can, e.g., be based on the inverse Rosenblatt transformation for cases where the joint density function of the variables is formulated by means of the conditional modelling approach [21] or by the Nataf transformation if the marginal PDFs of variables and the correlation coefficients between these variables are given [12]. The environmental contour obtained by the IFORM is accordingly the collection of physical environmental parameters that correspond to the normalized values, which are located on the sphere with radius β_F in the U space.

It is found that the main benefit of the design contour method is due to the description of the environmental parameters being uncoupled from computation of the structural response. Due to its high efficiency and satisfactory accuracy, the environmental contour method has, e.g., been widely applied for assessment of ULS criteria, and in particular during preliminary design phases. The most critical response corresponding to the environmental conditions defined by the environmental contour can then be applied for estimation of the extreme response for the specified return period. If the response process for a given ice condition is stochastic, the extreme response can be taken to be an upper fractile of the corresponding short-term extreme response distribution. This is similar to the approach applied in the case of wave-induced response [11].

4. First-Year Ice Ridges

4.1. Properties of Sea Ice Ridges

Sea ice that has survived one or more summers is referred to as old ice or multiyear ice. Typically, the relevant mechanical properties are then stronger than those corresponding to first-year sea ice [22]. It is pointed out in [23] that old ice could be encountered by Arctic ships and accordingly poses a threat to the ships. However, in the present study, only first-year sea ice ridges are considered. For one thing, the physical and mechanical properties of the first-year ice have been studied rather extensively, while very limited information is available with respect to old ice. The mechanical properties are assumed to be close to those of the surrounding level ice. Furthermore, most of the old ice is found at very high latitudes and the ice conditions, e.g., for the typical NSR are mostly first-year. Ice ridges are regarded as one of the major hazards that need to be considered for Arctic shipping routes. Ice ridge is a line or wall of broken ice components that are forced upwards due to shear or pressure. At an early stage, the ice ridges are formed as broken ice rubble. These blocks may subsequently to some extent become consolidated at the refreezing stage. As illustrated in Figure 2, there are generally three distinct parts in a first-year ridge [3]: the above-water sail part, the consolidated layer, and the ice rubble. The sail part has pores filled with air and snow. The consolidated layer forms the upper part of the ridge keel and is a continuous layer of ice. The ice blocks found in the lower part of the keel are loose and only partially refrozen. Due to the buoyancy forces, these rubble blocks are packed together with water being trapped between the blocks. Figure 3 gives an illustration of some of the key parameters that can characterize the geometry of a first-year ridge.

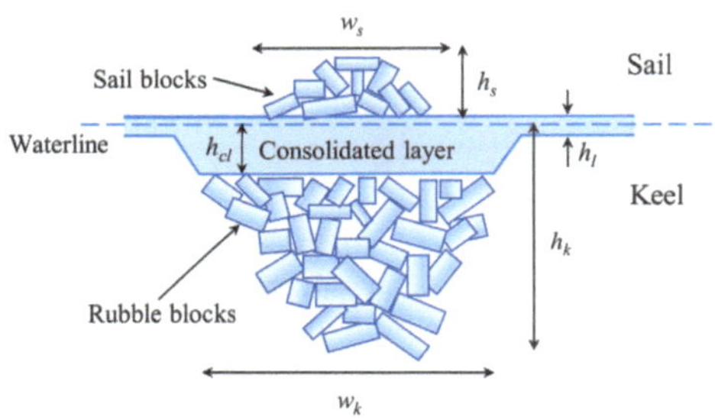

Figure 3. First-year ice ridge with the key parameters: sail width w_s, sail draft h_s, consolidated layer thickness h_{cl}, level ice thickness h_l, keel width w_k, and keel draft h_k [24].

4.2. Ship–Ridge Interaction

A number of studies related to the interaction between ships and level ice, and ships and ice floes have been performed, e.g., [25–28]. However, there are few studies related to ship–ridge interaction, which is most likely due the complexity of such interaction processes. There are basically three different methods for prediction of structural loads caused by ridges, i.e., analytical (or semi-analytical) methods, numerical simulation models, and experimental tests. For the analytical and semi-analytical methods, the ship and ice interaction model is much simplified and well-suited for fast calculations. As an example, the loads caused by the sail can be neglected because its size is small compared to that of the keel [3]. Ice forces acting on ship-shaped structures can be divided into main load components–those caused by the consolidated layer part versus ones caused by the rubble blocks. Generally, the consolidated layer is 2.0–2.5 times thicker than the surrounding level ice. Furthermore, the mechanical properties of the consolidated layer are assumed to be close to those of the surrounding level ice. Therefore, the consolidated layer part can be considered as a thick level ice and the corresponding load components acting on fixed structures can be estimated by application of relevant formulas recommended in ISO 19906. For the load component caused by the rubble blocks, another empirical formula based on a passive mode of failure is also found in the same document.

In numerical simulation models, the consolidated layer and the rubble are in many cases modeled respectively as a thick level ice and a granular material. A number of numerical studies have been performed to study the interaction between a ship and level ice. An example is the three degree-of-freedom (3DOF) rigid model developed in [27]. This model has later been extended to comprise six degrees-of-freedom (6DOF) by Tan et al. [28]. Other studies are, e.g., those based on the finite element method (FEM), the graphics processing unit (GPU) computation basis [2,29], and the cohesive element method (CEM) [30]. In order to study the interaction between ice rubble and ships, numerical methods such as the DEM [31,32] and the FEM [33] have been applied. However, due to the complexity of the ice ridge and ship interaction process, little work has been carried out in relation to numerical simulation of the entire process where the simultaneous loading caused by both of the two different keel components have been accounted for.

Improved understanding of ridge breaking processes can be achieved by means of scaled model tests and field tests, but only results from a limited number of such studies are available. Model tests related to interaction between ships and ice ridges were, e.g., performed in the Hamburg Ship Model Basin (HSVA) [34]. The ridge sail as well as the keel (including both a consolidated layer and ice rubble) were represented. A ship can break through a ridge either in a continuous manner or by means of ramming. The type of interaction depends on the ridge breaking energy, and the propulsive and the kinetic energy of the ship. Model tests that were carried out at HSVA for a moored floater having a sloping surface have confirmed that ridge ice loads are significantly higher than the loads due to the adjacent level ice [35]. Field experiments related to ship hulls interacting with first-year ridges have also been reported, e.g., [36–38].

Based on the abovementioned studies of first-year ice ridges interacting with ships and sloping structures, the ship and ridge interaction process is illustrated in Figure 4. During early design stages, loading due to the ridge sail part can be neglected. Furthermore, ice loads due to the underwater keel can be divided into two parts: the component caused by the consolidated layer and the component due to the unconsolidated ice rubbles. Ice loads from the consolidated layer part are assumed to correspond to that due to thick level ice. The ship and level ice interaction process is generally initiated by a localized crushing of the ice edge. Subsequently, the contact area and the crushing force will increase when the ship penetrates the ice feature. The ice sheet eventually deforms out of its plane, and the resulting bending stresses induce a flexural failure at a certain distance away from the region where crushing occurs [5]. The rubble can be broken by means of the kinetic and propulsive ship hull energy, and by the broken parts of the consolidated layer. The local water current may also serve to clear the rubble and the broken ice pieces from the keel.

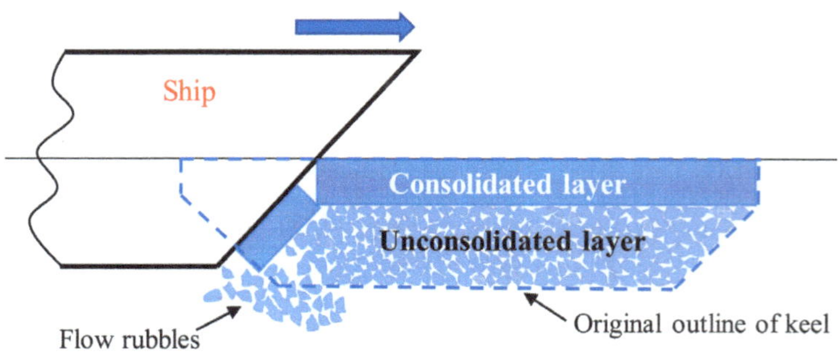

Figure 4. Illustration of a ship and ice ridge interaction process.

4.3. Statistical Models for Key Parameters Associated with Ice Ridges

The description of the ship–ridge interaction process given above gives reason to select four variables for characterization of ice ridges, and these also form the basis for establishment of environmental design contours. These four variables are the thickness of the consolidated layer, its crushing strength, its flexural strength, and the draft of the keel. Statistical models are introduced for each of these variables.

The thickness of the consolidated layer depends on geographical area, its environmental conditions, and the time of the season. Here, the consolidated layer thickness of ridges located in the Barents Sea is considered. The data sets for the thickness of the consolidated layer were collected from field measurements that were carried out from 2002 to 2011 [24].

Based on a statistical analysis, the Gamma distribution is found to provide a satisfactory description of the experimental data. The Gamma probability density function is given by:

$$f(h_{cl}) = \frac{1}{\Gamma(k)\theta^k} h_{cl}^{k-1} exp(-\frac{h_{cl}}{\theta}) \tag{7}$$

where $\Gamma(\cdot)$ represents the gamma function, and θ and k are the scale and the shape parameters, respectively. The resulting values of the two parameters are found to be 2.97 (shape) and 0.54 (scale) based on application of the moment estimators. Fitting of a Gamma distribution to the data is illustrated in Figure 5 [39].

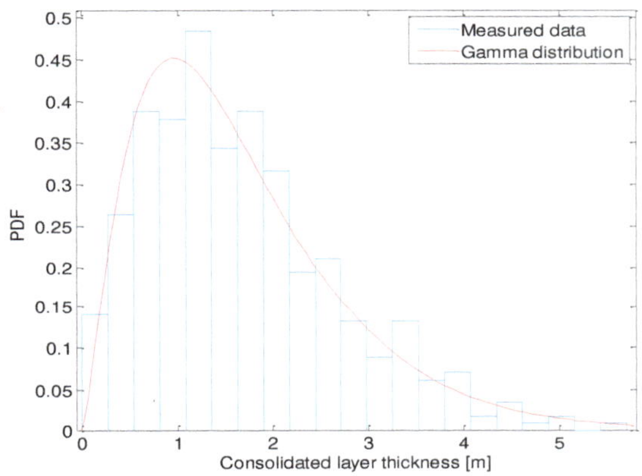

Figure 5. Density function for the thickness of the consolidated layer h_{cl} fitted to the measurements.

The distribution of the consolidated layer thickness will be different for different Polar regions, and also associated with large uncertainties. The data measured in the Barents Sea were obtained based on mechanical or thermal drilling, and these provide at least a certain amount of information. For other Polar regions, where only limited or lacking data coverage exists, utilization of data related to the thickness of the surrounding level ice can possibly provide an estimate of the distribution of the consolidated layer thickness by application of a representative "enhancement factor".

Because there are very limited experimental data for the mechanical properties of the consolidated layer, these values are assumed to be close to those of the surrounding level ice due to limited experimental data for the consolidated layer of ice ridges. For the flexural strength of first-year sea ice, a large number of measurements have been performed in different Arctic regions [22]. This strength parameter depends basically on the ice temperature, salinity, and inherent brine volume. The data of flexural strength in the present study are based on those provided by Timco [40]. These comprise in situ experimental data obtained from different Arctic regions, such as Baffin Island, Greenland, and the Gulf of Bothnia, and the Canadian and Alaskan Beaufort Sea. The two-parameter Weibull distribution can typically be applied to describe probabilistic distribution of the flexural strength σ_f, which is expressed as:

$$f\left(\sigma_f\right) = \frac{\beta}{\alpha}\left(\frac{\sigma_f}{\alpha}\right)^{\beta-1} exp\left(-\left(\frac{\sigma_f}{\alpha}\right)^{\beta}\right) \tag{8}$$

where the scale parameter α and the shape parameter β are determined as 0.582 and 2.090, respectively, by means of regression estimators as implemented in a probability paper. A fitted Weibull density function (PDF) for the flexural strength of first-year sea ice in the Arctic is shown in Figure 6 [39].

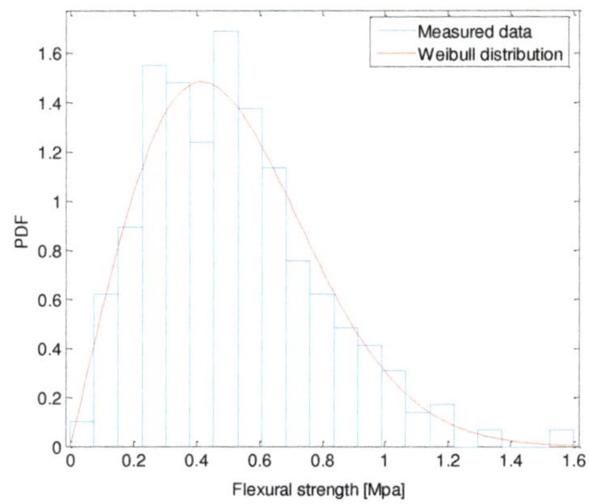

Figure 6. Density function for the flexural strength for first-year level ice σ_f, fitted to measurements.

The level ice crushing strength depends on various parameters such as loading direction, porosity, salinity, temperature, failure mechanism, etc. For ship–ridge interaction, samples that are loaded in the horizontal direction should be considered. However, only limited data are published for such samples in relation to Arctic regions. An exception is the data from experiments in the Svalbard region and the Barents Sea. Based on experiments for the winter seasons during 2005–2012 [41,42], data from 363 horizontally loaded samples were collected. Based on a statistical analysis, it is found that the lognormal distribution, given by Equation (9), can provide a satisfactory description of the measurements.

$$f(\sigma_c) = \frac{1}{\sqrt{2\pi}\sigma \cdot \sigma_c} exp(-\frac{(ln\sigma_c - \mu)^2}{2\sigma^2}) \tag{9}$$

Here, σ_c represents the crushing strength of the ice feature, while μ = 0.644 and σ = 0.550 are, respectively, the mean value and the standard deviation (of the logarithmic values).

The fitted lognormal density function and histograms of the measured data are shown in Figure 7 [39]. It should be noted that the present probabilistic model for the crushing strength is based on the measured samples from the Barents Sea and Svalbard regions. This model can also be applied for representation of the distribution of crushing strength for first-year sea ice in other Arctic regions, but relevant data and experiments from other regions are also required in order to enrich the current probabilistic model.

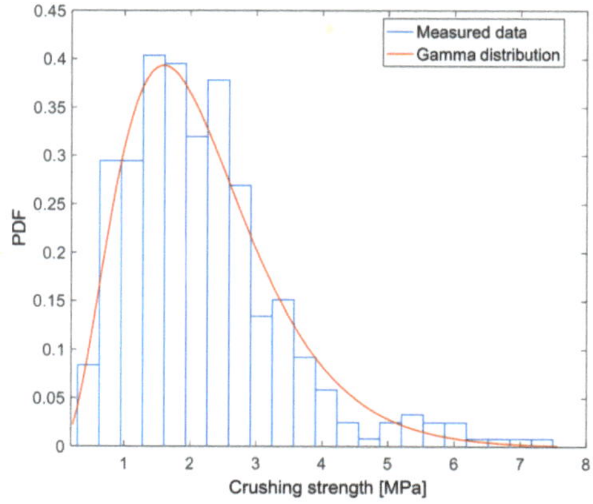

Figure 7. Fit of PDF for the first-year level ice crushing strength $\sigma_{c,}$, versus measured data.

Ridge loads acting on ship hulls due to the rubble blocks depend on the keel draft, the keel width, the keel porosity, and the mass of the keel part. Among these parameters, the draft of the keel h_k is the most prominent. This quantity depends on geographical location and time of the season, similar to the consolidated layer thickness, e.g., Samardžija and Høyland [43].

The magnitudes of the draft of the keel can be obtained by means of drilling or by performing continuous scanning. Data for the keel draft of first-year ridges in Arctic regions such as the Barents Sea, the Greenland area, and the Beaufort Sea, which are subjected to statistical analysis, form the basis for the present study. These data sets are published in [5], and they are mainly collected by mechanical and thermal drilling. It is observed that the exponential model with a PDF expressed as follows can give an adequate fit to the measurements [44]:

$$f(h_k) = \lambda exp(-\lambda(h_k - c)) \tag{10}$$

Here, h_k denotes the draft of the keel and λ is the parameter of the exponential model. The constant c can be regarded as a lower threshold (i.e., a "cut-off value") for the keel draft.

When the data are obtained from a collection of different Polar regions, however, the three-parameter Weibull model given by Equation (11) is found to provide a better fit to the collective data sets than what can be achieved by means of the exponential distribution:

$$f(h_k) = \frac{\beta}{\alpha}\left(\frac{h_k - c}{\alpha}\right)^{\beta-1} exp\left(-\left(\frac{h_k - c}{\alpha}\right)^{\beta}\right) \tag{11}$$

Here, the scale parameter is estimated as α = 9.464, the shape parameter β is estimated as 1.697, and the location parameter is estimated as c = 0.450, based on application of the least square method.

Note that seasonal variations of the parameters that are applied for characterization of ice ridges have presently not been considered. Recent studies of such effects can be found, e.g., in [43,45,46].

5. Environmental Design Contours for Ice Ridges

The key ice ridge parameters and the associated probabilistic models pertaining to these parameters are described above. Based on these probabilistic models and the IFORM approach, various categories of environmental contours can be developed. This comprises two-dimensional contour lines, three-dimensional contour surfaces, and four-dimensional manifolds. These can subsequently be employed within the context of deterministic or reliability-based design of ships in Polar waters. Different types of such environmental contours are discussed in this section.

5.1. Two-Dimensional Contours

Among the three forms of environmental contours mentioned above, the environmental contour lines based on only two environmental parameters represent the simplest form. The ship–ridge interaction process can then be represented in a simplified manner as a sloping structure exposed to an incoming ice sheet. Ship-ice loads can accordingly be estimated by applying the empirical formula given in [3], which is used to calculate the static ice loads. This formula only requires the flexural strength and the consolidated layer thickness of the ice feature as input quantities. This simple model is clearly not completely satisfactory for representation of the ship–ice ridge interaction process, since both dynamic effects as well as ridge keel actions are not taken into account.

Still, such a model can be applied as a basis for illustration of the main steps associated with application of the design contours for estimation of the extreme ice ridge loads. Moreover, the effect of increasing correlation between the basic design parameters and an increasing number of encountered ice ridges corresponding to a given return period in relation to the resulting environmental contours can readily be investigated, which is due to the simplicity of the two-dimensional design contours.

The specific location that is considered determines the probability distribution of the consolidated layer thickness. Presently, only experimental data that are collected for h_{cl} in the Barents Sea are available. Accordingly, it is assumed that the relevant Arctic ship mainly operates in the Barents Sea, and, furthermore, that it travels a distance of 5000 km per year in areas with ice ridges. The density of ice ridges is taken to be 2/km along the route [47]. The intended lifetime is assumed to be 50 years. Presently, ice conditions along the route are assumed to be constant during this period.

The annual number of ice ridges encountered by the ship is assumed to be a fixed value, which is given as $N_{1year} = r{\cdot}5000$ km$\cdot$2/km, where r is the encounter frequency that depends on the capability of the navigation equipment on board and the experience of the crew. The number of ridges experienced by the ship hull during the service life, $N_{50years}$, is then obtained:

$$N_{50years} = 50 \cdot N_{1year} = 50 \cdot r \cdot 5000 \cdot 2 \tag{12}$$

The radius of the circle in the normalized space, β_F, which corresponds to this return period is determined by:

$$\beta_F = \Phi^{-1}\left(1 - p_f(y_N)\right) = \Phi^{-1}\left(1 - \frac{1}{N_{50years}}\right) \tag{13}$$

A circle corresponding to r = 0.5 (which gives $N_{50years}$ = 250,000) is plotted in Figure 8a, where u_1 and u_2 are independent normalized Gaussian variables given by $u_1 = \beta_F{\cdot}\sin(\eta)$ and $u_2 = \beta_F{\cdot}\cos(\eta)$, where the angle η ranges between 0 and 2π.

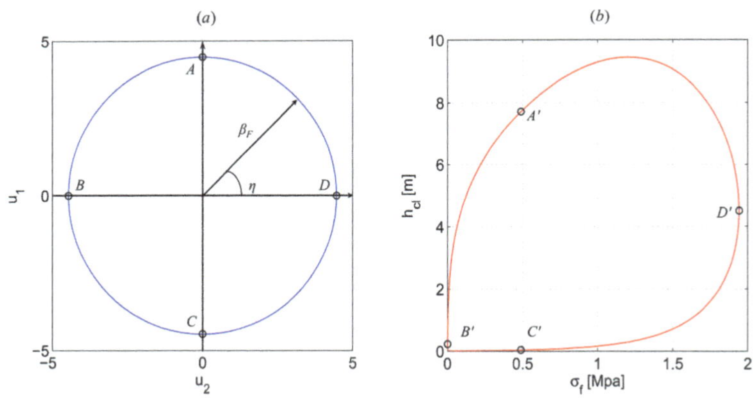

Figure 8. The circle in normalized space with radius β_F (**a**) and the resulting design contour line in the physical domain (**b**) for a case with $N_{50years}$ = 250,000.

Presently, only the marginal PDFs of the key parameters are available, while the correlation coefficients between the two parameters are not known. The variables s_1 and s_2 are introduced, which correspond to the consolidated layer thickness and the ice crushing strength, respectively. The resulting design contour in physical parameter space is then established based on the circle in normalized space by application of the Nataf transformation, which is expressed as follows:

$$s_1 = F_{S_1}^{-1}(\Phi(u_1))s_2 = F_{S_2}^{-1}\left(\Phi\left(u_2\sqrt{1-\rho'^2_{12}} + \rho'_{12}\Phi^{-1}\left(F_{S_1}(s_1)\right)\right)\right) \tag{14}$$

Here, ρ_{12} denotes the correlation coefficient between the crushing strength and the thickness of the consolidated layer. This coefficient is related to ρ'_{12}, which denotes the associated (equivalent) coefficient of correlation that is applied by the Nataf transformation. The two correlation coefficients are connected by a semiempirical equation of the following type. The structural response corresponds to

$$\rho'_{12} = \zeta \cdot \rho_{12} \tag{15}$$

where relevant expressions for the function ζ can be found in [48].

The contour line for a 50-year return period is presented in Figure 8b for the case that the correlation coefficient ρ_{12} is 0.5. In order to illustrate the result of the transformation into physical space, four points that are located on the circle in normalized space are selected. These are denoted by A, B, C, and D in the figure. The four points are then mapped into the corresponding points in the physical parameter space. These are designated by A', B', C', and D'. Having generated the contour line corresponding to a 50-year return period, the associated extreme load level, y_N, can be estimated based on the following principle:

$$y_N \approx \text{maximum load around the } (s_1, s_2) \text{ contour} \tag{16}$$

Accordingly, the extreme response/loads can be simplified by searching along the contour for the most critical environmental condition. The values of correlation coefficient ρ_{12} and the encounter frequency r are selected somewhat arbitrarily due to the limitation of reference data. The influence of these two variables on the subsequent contour lines can be studied. For a given value of r, the effect of increasing the correlation coefficient ρ_{12} on the environmental contour is first shown in Figure 9.

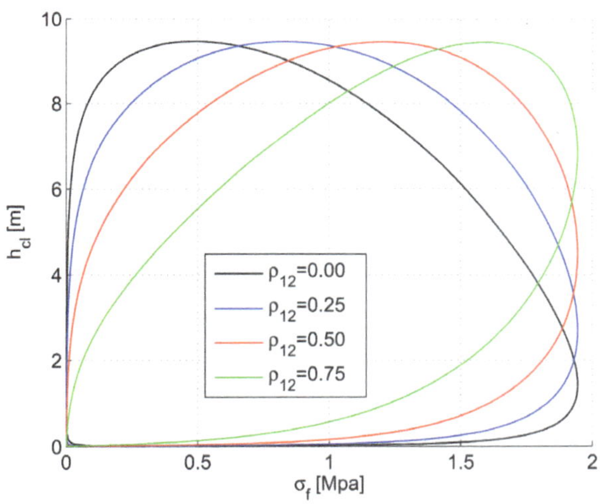

Figure 9. Effect on the shape of the contour lines caused by increasing correlation between the flexural strength and the consolidated layer thickness (for a specific value of the return period, which is N = 50 years for the present case).

From this figure, it is seen that the shapes of the environmental contour lines are strongly influenced by the value of the correlation coefficient. There are no changes of the maximum values for the flexural strength and the consolidated layer thickness along the different contour lines for varying values of ρ_{12}. However, the critical region where the simultaneous values of the consolidated layer thickness and the flexural strength are high is seen to become increasingly narrow for increasing values of the correlation coefficient. This narrowing implies that high values of consolidated layer thickness have a stronger tendency to be associated with high values of flexural strength, and this generally implies more serious ice loads. Increasing values of the correlation coefficient will accordingly generally imply an increase of the maximum loads caused by the ice conditions along the contour line (which also implies an increase of the extreme hull response associated with a given return period).

The influence of encounter frequency r on the environmental contour is next studied for the case that the correlation coefficient has a specific value of 0.50. The contour lines corresponding to varying r values then have similar shapes since they are based on the same coefficient of correlation. Increased value of r implies that a higher number of load events are experienced by the ship hull for the same return period. The corresponding contour lines then expand accordingly. The maximum value along the thickness axis changes from 9 m to almost 10 m when r changes from 0.2 to 0.75. For the flexural strength, the maxim value changes from 1.8 MPa to almost 2 MPa. Clearly, this gives increasingly higher ridge loads as the encounter frequency increases.

For stationary structures (e.g., floating production systems, FPSOs), the ridge encounter frequencies depend upon the arrival rate at the site where the structure is located. Accordingly, it cannot be influenced or controlled unless, e.g., disconnect systems or protective barriers are applied. An intermediate situation becomes relevant for floating units that are intended for temporary but extended operations, such as exploration, drilling, and installation vessels. An example of computed deformations and stresses caused by the impact of an ice ridge on the upward sloping hull of a mobile drilling unit (MODU) is shown in Figure 10 [49]. The structural response corresponds to permanent plastic deformations without any fracture taking place, and accordingly this would correspond to a mechanical limit state of the ULS/ALS category.

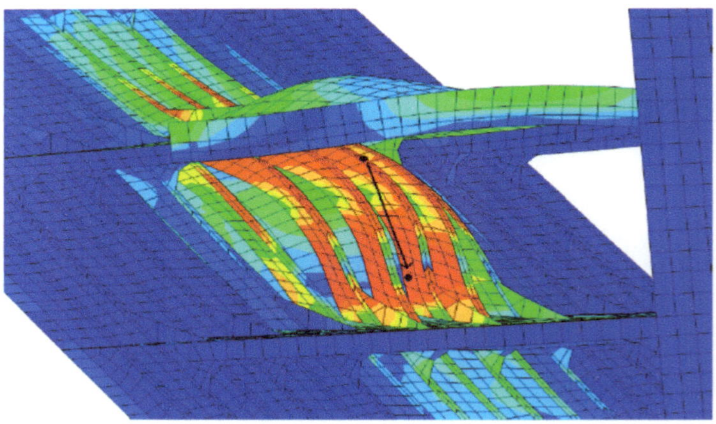

Figure 10. Example of computed deformations and von Mises stress levels caused by the impact of an ice ridge on the upward sloping hull of a mobile drilling unit (MODU).

5.2. Three-Dimensional Contour Surfaces

In order to obtain more general types of ice-loads than those restricted to the two-parameter model proposed in Section 5.1, models with three parameters for characterization of the ice ridge properties can also readily be accommodated within the present approach. Two specific examples of three-parameter combinations are (i) layer thickness/crushing strength/flexural strength and (ii) layer thickness/keel draft/flexural strength. Both of these cases are considered in [39]. Here, only the first alternative is addressed.

For this model, possible load contributions caused by the unconsolidated rubble are excluded. The main ice loads caused by the ship and ridge interaction are now assumed to be dominated by the consolidated layer part [14]. However, dynamic effects associated with the ship–ice interaction process can be included as part of numerical simulations or relevant empirical or theoretical load formulations. Accordingly, the consolidated layer thickness, the flexural strength, and the crushing strength are included as the relevant parameters. As compared to the contour based on two variables, the third component s_3

represents the ice ridge crushing strength. The corresponding variable u_3 in the normalized U space is also now expressed by means of the Nataf model based on introduction of two additional equivalent correlation coefficients (i.e., ρ'_{13} and ρ'_{23}):

$$\begin{aligned} s_1 &= F_{S_1}^{-1}(\Phi(u_1)) \\ s_2 &= F_{S_2}^{-1}\left(\Phi\left(u_2\sqrt{1-\rho'^2_{12}}+\rho'_{12}\Phi^{-1}(F_{S_1}(s_1))\right)\right) \\ s_3 &= F_{S_3}^{-1}(\Phi(\frac{u_3}{\sqrt{1-\rho'^2_{12}}}\sqrt{1-\rho'^2_{12}-\rho'^2_{13}-\rho'^2_{23}+2\rho'_{12}\rho'_{13}\rho'_{23}} \\ &+\frac{1}{1-\rho'^2_{12}}((\rho'_{13}-\rho'_{12}\rho'_{23})\Phi^{-1}(F_{S_1}(s_1))+(\rho'_{23}-\rho'_{12}\rho'_{13})\Phi^{-1}(F_{S_2}(s_2))) \end{aligned} \tag{17}$$

where the quantities ρ'_{ij} $(i, j = 1, 2, 3; i \neq j)$ designate equivalent correlation coefficients that are applied by the Nataf transformation.

As an example, a case is considered for which all the three correlation coefficients in the joint statistical mode are equal to 0.5. The encounter frequency r is assumed to be 0.5 and the parameters of the statistical models are set to be the same as those described above. As an extension of the two-dimensional contours, a sphere with radius β_F in three dimensions is next established in the normalized space. A three-dimensional design contour in the physical parameter space is subsequently obtained by means of a transformation based on the Nataf model according to Equation (17). The resulting contour surface with three key parameters, i.e., the consolidated layer thickness, the flexural strength, and the crushing strength (corresponding to a return period of 50 years) is plotted in Figure 11 [39].

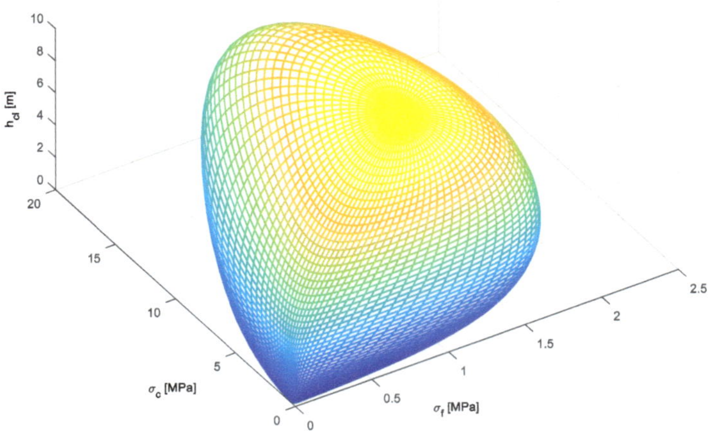

Figure 11. The contour surface based on three parameters: thickness of consolidated layer, flexural strength, and crushing strength, for a return period of 50 years.

To provide a detailed visualization of the contour surface, a suite of contour lines in two dimensions, which are established by locking the value of one of the parameters (e.g., the consolidated layer thickness), could have also been constructed.

Having established the contour surface that corresponds to the three basic parameters selected, a set of ice ridge conditions that correspond to specific joint values of these parameters can be selected. These conditions correspond to specific points located on the contour surface. For each of these points, a "response analysis" is performed, e.g., by means of numerical calculations, empirical expressions, analytical solutions, and/or experimental tests. The most critical ice ridge condition is then subsequently identified. This yields an estimate of the highest ice ridge load and allows calculation of the associated extreme 50-year hull response.

5.3. Environmental Contour for the Case with Four Parameters

Although the model just discussed with three basic ice ridge parameters is more comprehensive than the model with only two parameters, it may still not be adequate for some design purposes. As an example, the possible contribution to the ice loading caused by the unconsolidated rubble blocks is not considered. Accordingly, an even more refined four-parameter model, which also includes the keel draft as an additional random variable, can be relevant. Based on such a statistical representation, both the dynamic effects associated with the interaction between the consolidated layer and the ship hull, and the load effects due to the rubble blocks can be taken into account.

It is relevant to assume that there is no correlation between the keel draft and the other three key parameters. The other coefficients of correlation ρ_{ij} (with $i, j = 1, 2, 3$; and $i \neq j$) are hence set equal to 0.5 for the purpose of illustrating the resulting contour. The other parameters, such as r and $N_{50years}$, are kept the same as for the example calculations above. For the purpose of visualizing the four-parameter contour (for a return period of 50 years), a set of environmental contour surfaces which correspond to specific values of the keel draft are displayed in Figure 12.

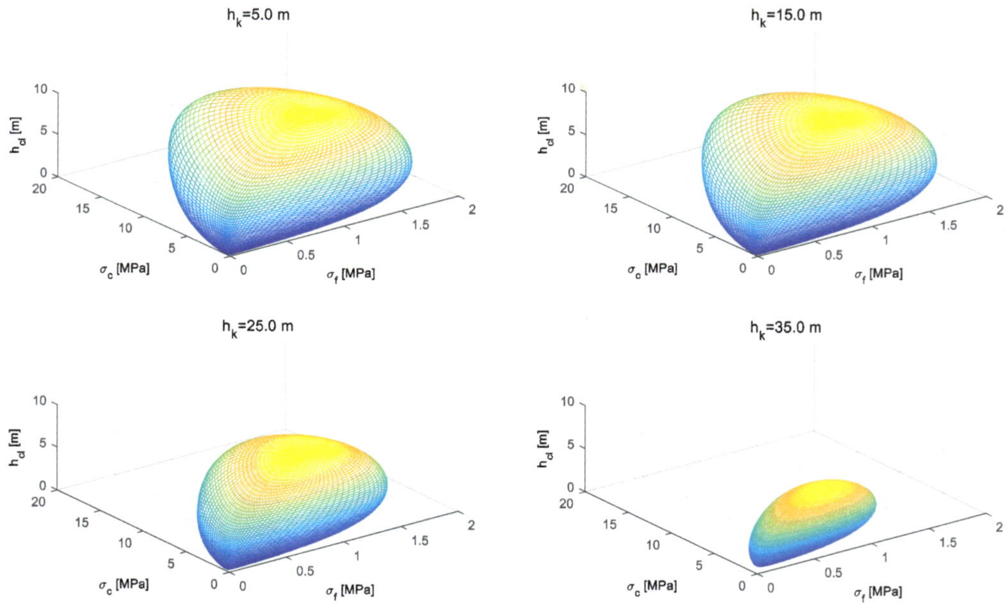

Figure 12. Three-dimensional 50-year contours that represent level surfaces corresponding to different keel drafts h_k.

For a specific keel draft, the equivalent value of u_4 in normalized space is obtained by applying the transformation given by Equation (17). Then, a sphere of radius β_F (given in Equation (13)) is established in the normalized space of three standard Gaussian variables u_1, u_2, and u_3. This three-dimensional sphere is subsequently transformed into the space of physical parameters that results in the contour surfaces shown in Figure 12. The surfaces that correspond to different values of h_k are of a similar shape due to identical correlation relationships with respect to the four basic parameters. It is also seen that for keel drafts above the mean value (h_k = 8.89 m), the volume that is bounded by the contour surface decreases for increasing values of the draft. Having established such a set of contour surfaces associated with the return period of 50 years, the next task is for the designer to find the ice ridge conditions that are located on the surface (for a given draft), which cause the highest response levels in the ship hull. This requires consideration of a range of keel drafts, which adds some effort for identifying the most critical ridge characteristics, as compared to the case with only three parameters.

6. Design Contours for Growlers, Bergy Bits, and Icebergs

6.1. General

There are two mainly different populations of ice features of this type: (i) Growlers and bergy bits with water lengths of the order of 0–5 m and 5–20 m, respectively. The corresponding mass intervals are from 0 to 1000 tons for growlers and from 1000 tons to 10,000 tons for bergy bits, according to Raghuvanshi and Ehlers (2015). (ii) Icebergs classified are into *small icebergs* from 10,000 tons to 100,000 tons, *medium icebergs* from 100,000 tons to 1,000,000 tons (1 million tons), and *large icebergs* from 1,000,000 tons and greater [50].

The probability distribution for the mass corresponding to each of the populations is assumed to be of the exponential type. According to Fuglem et al. [51], the mass of both populations can be related to the waterline length according to the following formula:

$$M_{Iceberg} = 0.3\, L_w^3 \text{ (length in m and mass in tons)} \tag{18}$$

A somewhat more refined formula involving the dimensions along the three orthogonal directions is given in [50].

6.2. Two-Dimensional Design Contours

The PDF for the water line length proposed in [51] is expressed as:

$$f_{Lw}(x) = 0.43\left(\frac{1}{60}e^{-\frac{x}{60}}\right) + 0.57\left(\frac{1}{8}e^{-\frac{x}{8}}\right) \tag{19}$$

This is due to the hypothesis that the number of icebergs with L greater than 20 m should approximately be equal to the number of icebergs with waterline length between 5 and 20 m. In the following, this probability model is applied for the purpose of illustrating the construction of two-dimensional environmental contours by also including the iceberg drift velocity as a random variable.

Data of ice drift speed in the Pechora Sea were collected by Løset and Onshus [48]. They found that a two-parameter Weibull distribution gave an adequate fit to the measurements:

$$F_X(x) = 1 - \exp\{-(x/\theta)^\gamma\} \tag{20}$$

where the scale and shape parameters are obtained as $\theta = 0.213$ and $\gamma = 1.406$, respectively.

The probability for a ship to experience an impact from an ice feature is assumed to be higher for the smallest ice features (growlers and bergy bits) versus larger icebergs (given that a geographical encounter with such an object has taken place). Denoting the occurrence rates for the two main categories of ice features by ρ_1 and ρ_2 (per traveled distance unit for the ship), and the corresponding conditional collision probabilities by p_{coll1} and p_{coll2}, the numbers of impacts during a period of N_{year} are then obtained as follows (with subscript 1 denoting smaller ice features and subscript 2 referring to larger features):

$$N_1 = p_{coll1} \cdot N_{year} \cdot \rho_1 \cdot L N_2 = p_{coll2} \cdot N_{year} \cdot \rho_2 \cdot L \tag{21}$$

where L is the average distance traveled by the ship in Polar waters per year. The corresponding probabilities of exceedance and reliability indices are then obtained as:

$$P_{f1} = \frac{1}{N_1}, \beta_{F1} = \Phi^{-1}\left(1 - P_{f1}\right) P_{f2} = \frac{1}{N_1}, \beta_{F2} = \Phi^{-1}\left(1 - P_{f2}\right) \tag{22}$$

By application of the cumulative distribution for the mass of the ice feature and the cumulative distribution for the drift speed, the corresponding two-dimensional environmental design contour can be obtained (by also assuming that mass and drift speed are uncorrelated). It is reasonable to consider collision with the two main categories of ice features as two distinctly different events, since the induced structural damage level typically is significantly different. Accordingly, the cumulative distribution function for the mass is also split in two by application of the respective terms for the density function in Equation (19).

As an example, a case where the number of impact events for the growler/bergy bit category is N_1 = 750 is considered, which implies that the corresponding exceedance probability is 0.00135. This gives a circle with radius equal to three in the normalized plane, and the resulting mass–velocity contour in the physical parameter plane is shown in Figure 13.

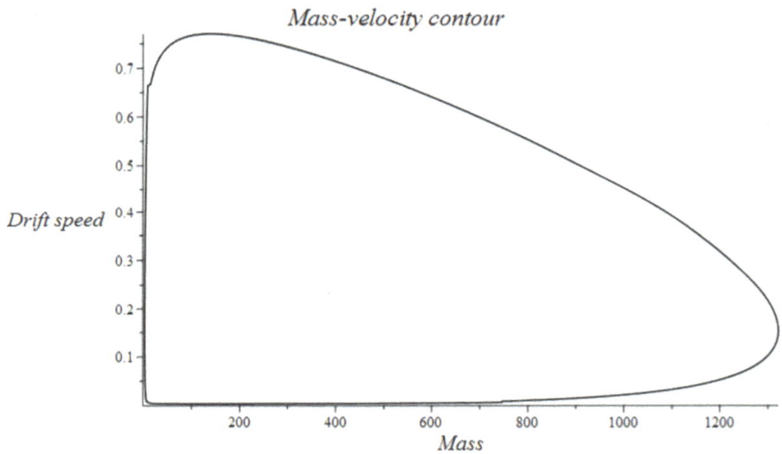

Figure 13. Mass–velocity contour for growlers/bergy bits corresponding to N_1 = 750.

By also including the ship speed as a deterministic parameter, a sequence of two-dimensional contours of a similar type is obtained. These are shown as a 3D surface in Figure 14a with the relative speed (ship speed plus drift speed) along the second horizontal axis, and the drift speed along the vertical axis. The corresponding 2D projections are illustrated in Figure 14b.

In order to identify the relevant design scenario in terms of ice feature mass and relative velocity, the level curves corresponding to constant kinetic energy are required. The corresponding tangent point between these level curves and the mass–velocity contours then represent the most critical scenario along the environmental contour. This is illustrated in Figure 15. It is seen that for increasing ship speeds, the tangent points are shifted towards increasingly greater values for the mass of the ice feature.

Examples of numerically simulated structural damage for this category of ice features are shown in Figure 16 [18]. In the left part, Figure 16a shows the damage level in terms of permanent plastic deformations that would correspond to ULS/ALS-type of design criteria. In the right part, Figure 16b illustrates a damage that resulted in fracture of the steel plating, and this would correspond to an ALS-type of criterion.

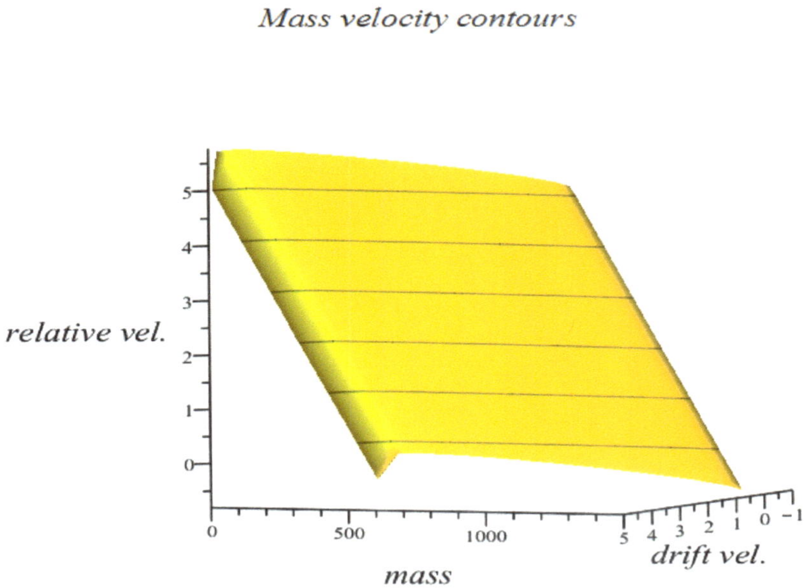

(**a**)

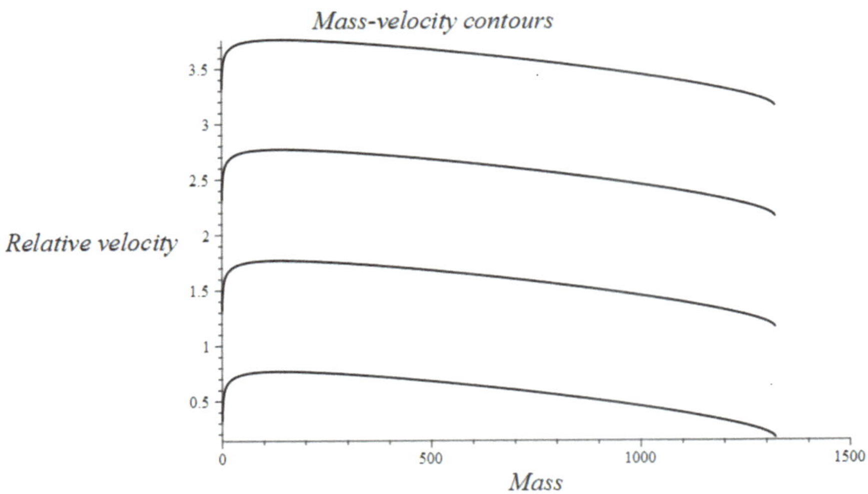

(**b**)

Figure 14. Sequence of 2D mass–velocity contours for increasing (deterministic) ship velocity. (**a**) 3D surface and (**b**) 2D projections.

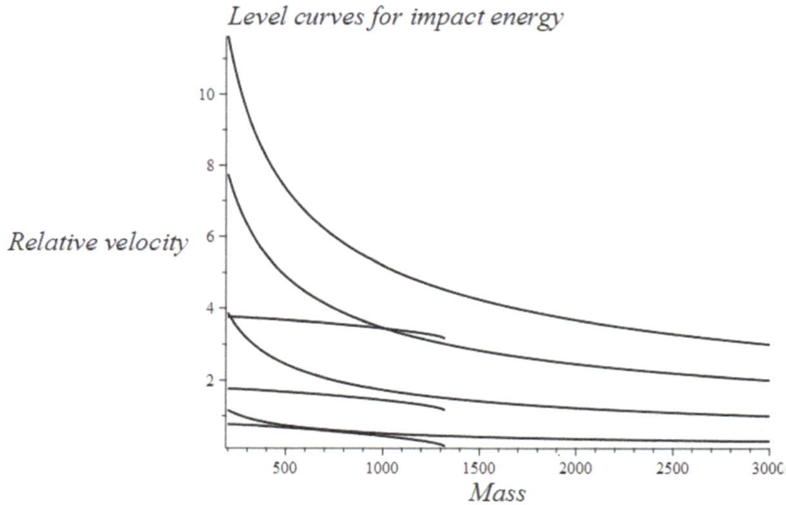

Figure 15. Level curves of constant kinetic energy versus mass–velocity contours.

(**a**)

(**b**)

Figure 16. Examples of simulated damage caused by impacting ice floe/ridge. (**a**)ULS/ALS-type of damage and (**b**) ALS-type of damage.

6.3. Inclusion of Additional Iceberg Collision Parameters

There are a number of additional parameters that influence the amount of energy absorbed by the ship hull. Examples include local shape of the ice feature in the impact area, collision angle, material properties of the ship hull, elevation, and longitudinal position of impact location.

Some of these parameters can be characterized as belonging to the so-called "external impact mechanics", while others correspond to the "internal impact mechanics", e.g., [10,52–54]. However, little information is available in relation to statistics for the parameters that characterize the ice impact scenarios themselves. Possibly, relevant information could be extracted from collision statistics related to ship–ship interactions, e.g., [55].

It is found that the local shape of the ice feature in the impact area has strong influence on the amount of energy that is absorbed by the ship hull. Some examples of numerical models that are applied in order to represent various types of ice features are shown in Figure 17 [18]. It will be possible to include a "shape peakedness factor" as part of an extended three-dimensional environmental ice contour. A uniform probability model can initially be applied for such a parameter unless more specific information becomes available.

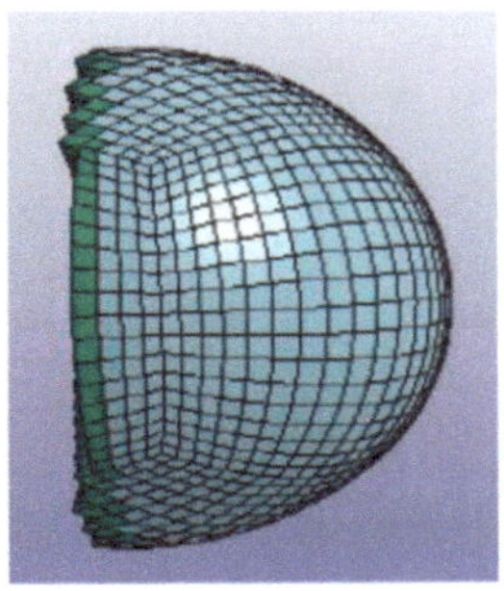

(**a**)

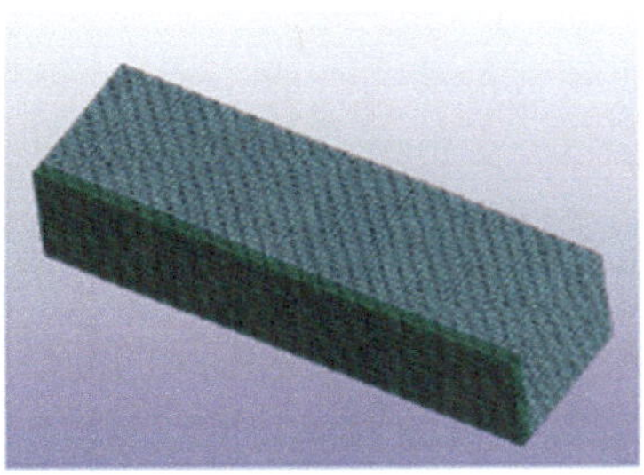

(**b**)

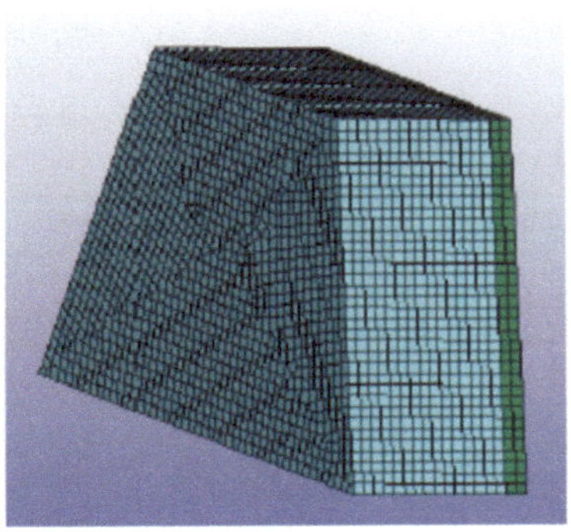

(**c**)

Figure 17. Examples of numerical models of ice features of different geometries: (**a**) growler, (**b**) bergy bit, and (**c**) iceberg.

7. Conclusions

In this work, based on the interaction processes between ships and ice ridges/icebergs, relevant parameters associated with these ice features for determining the loading on a ship hull are addressed. Probabilistic models are applied in order to describe the available data for key design parameters. The potential for application of environmental contours for analysis and design of ships in Polar regions is illustrated. By application of the inverse FORM method (IFORM), different categories of design contours are generated by reflecting the number of parameters associated with the process of ship–ice interaction.

The effect of increasing the correlation between the environmental parameters and the influence from the encounter frequency r on the resulting environmental contour are both studied. It is important that more research should be directed toward collecting data related to the degree of correlation among the physical/mechanical parameters of first-year sea ice since this has a strong effect on the design contour and on the extreme ice loads to which the ship hull is subjected.

The total number of ridge load events experienced by a Polar ship corresponding to a given return period is important for the resulting design load. The number of such events depends, e.g., on the frequency of encounter r, the annual travel distance, and the impact probability along the sailing route. In addition, global climate change also has some potential influence on the statistical characterization of the relevant parameters for the ice features considered. Such effects could be captured by a modified formulation of the proposed environmental contour method.

The data which form the basis for construction of the statistical models typically depend on area. Additional data for characterization of the level ice and the consolidated layer of ice ridges in other Polar areas are continuously being collected. Environmental contours for characterization of input parameters that are relevant for load and response calculations can accordingly be established also for these areas.

Furthermore, there is clearly a need for more explicit statistical information in relation to many of the other parameters that define relevant impact scenarios. As increasingly more empirical data become available, more accurate and refined design methods can be achieved.

This should also be seen in the light of recent design codes that move toward a more goal-based approach rather than the prescriptive rules of today [56–58]. This results in less specific details and less rigid design requirements. Design against accidental loads for ships in Polar regions must also take into account relevant ship operation criteria and corresponding transit restrictions for ships in Polar waters [59,60].

Author Contributions: Conceptualization, W.C., B.J.L., and G.R.; methodology, W.C., B.J.L., and G.R.; software, W.C., G.R., and B.J.L.; validation, W.C., G.R., and B.J.L.; resources, W.C., B.J.L., and G.R.; data curation, W.C.; writing—original draft preparation, W.C. and B.J.L.; writing—review and editing, W.C., B.J.L., and G.R.; visualization, W.C., G.R., and B.J.L.; supervision, B.J.L.; project administration, B.J.L.; funding acquisition, B.J.L. All authors have read and agreed to the published version of the manuscript.

Funding: This research was funded by the Lloyd's Register Foundation, a charitable foundation, helping to protect life and property by supporting engineering-related education, public engagement, and the application of research (www.lrfoundation.org.uk (accessed on 20 June 2021)). Some of this work has also been carried out within the MARTEC research project PRICE (Prediction of ice–ship–interaction for icebreaking vessels), and financial support from the Research Council of Norway (RCN project number: 249272/O80) is acknowledged.

Institutional Review Board Statement: Not applicable.

Informed Consent Statement: Not applicable.

Data Availability Statement: Not applicable.

Conflicts of Interest: The authors declare no conflict of interest. The funders had no role in the design of the study; in the collection, analyses, or interpretation of data; in the writing of the manuscript, or in the decision to publish the results.

References

1. Bergström, M.; Leira, B.J.; Kujala, P. Future Scenarios for Arctic Shipping. In *Proceedings of the Volume 7: Polar and Arctic Sciences and Technology*; ASME International: Washington, DC, USA, 2020.
2. Hahn, M.; Dankowski, H.; Ehlers, S.; Erceg, S.; Rung, T.; Huisman, M.; Sjöblom, H.; Leira, B.J.; Chai, W. Numerical prediction of ship-ice interaction: A project presentation. In Proceedings of the 36th International Conference on Ocean, Offshore and Arctic Engineering, Trondheim, Norway, 25–30 June 2017.
3. ISO. *ISO 19906: Petroleum and Natural Gas Industries–Arctic Offshore Structures*; ISO: Geneva, Switzerland, 2010.
4. Ehlers, S.; Cheng, F.; Jordaan, I.; Kuehnlein, W.; Kujala, P.; Luo, Y.; Freeman, R.; Riska, K.; Sirkar, J.; Oh, Y.-T.; et al. Towards mission-based structural design for arctic regions. *Ship Technol. Res.* **2017**, *64*, 115–128. [CrossRef]
5. Chai, W.; Leira, B.J.; Naess, A. Probabilistic methods for estimation of the extreme value statistics of ship ice loads. *Cold Reg. Sci. Technol.* **2018**, *146*, 87–97. [CrossRef]
6. Lensu, M. *Short Term Prediction of Ice Loads Experienced by Ice Going Ships*; Helsinki University of Technology: Espoo, Finland, 2002.
7. Riska, K. *On the Mechanics of the Ramming Interaction between a Ship and a Massive Ice Floe: Dissertation*; VTT Technical Research Centre of Finland: Espoo, Finland, 1987.
8. Riska, K.; Bridges, R. Limit state design and methodologies in ice class rules for ships and standards for Arctic offshore structures. *Mar. Struct.* **2019**, *63*, 462–479. [CrossRef]
9. Leira, B.J. A comparison of stochastic process models for definition of design contours. *Struct. Saf.* **2008**, *30*, 493–505. [CrossRef]
10. Yu, Z.; Amdahl, J. A numerical solver for coupled dynamic simulation of glacial ice impacts considering hydrody-namic-ice-structure interaction. *Ocean Eng.* **2021**, *226*, 108827. [CrossRef]
11. DNV GL. *Environmental Conditions and Environmental Loads (DNV-RP-C205)*; Det Norske Veritas AS: Oslo, Norway, 2014.
12. Silva-González, F.; Heredia-Zavoni, E.; Montes-Iturrizaga, R. Development of environmental contours using Nataf dis-tribution model. *Ocean Eng.* **2013**, *58*, 27–34. [CrossRef]
13. Winterstein, S.R.; Ude, T.C.; Cornell, C.A.; Bjerager, P.; Haver, S. Environmental parameters for extreme response: Inverse FORM with omission factors. In Proceedings of the ICOSSAR-93, Innsbruck, Austria, 9–13 August 1993; pp. 551–557.
14. Chai, W.; Leira, B.J. Environmental contours based on inverse SORM. *Mar. Struct.* **2018**, *60*, 34–51. [CrossRef]
15. Eckert-Gallup, A.C.; Sallaberry, C.J.; Dallman, A.R.; Neary, V.S. Application of principal component analysis (PCA) and improved joint probability distributions to the inverse first-order reliability method (I-FORM) for predicting extreme sea states. *Ocean Eng.* **2016**, *112*, 307–319. [CrossRef]
16. Lu, W.; Yu, Z.; Van den Berg, M.; Dennis, M.; Lubbad, R.; Hornnes, V.; Amdahl, J.; Løset, S.; Kim, E. Loads, Design and Operations of Floaters in the Arctic (Nord ST20). 2019. Available online: https://www.scinapse.io/papers/2905273220 (accessed on 21 June 2021).
17. Kim, S.J.; Korgersaar, M.; Ahmadi, N.; Taimuri, G.; Kujala, P.; Hirdaris, S. The influence of fluid structure interaction modelling on the dynamic response of ships subject to collision and grounding. *Mar. Struct.* **2021**, *75*, 102875. [CrossRef]
18. Radhakrishnan, G. Analysis of Accidental Iceberg Impacts with Large Passenger Vessels and FPSOs. Master's Thesis, NTNU, Trondheim, Norway, Aalto University, Espoo, Finland, 2018.
19. Madsen, H.O.; Krenk, S.; Lind, N.C. *Methods of Structural Safety*; Courier Corporation: Chelmsford, MA, USA, 2006.
20. Haver, S.; Winterstein, S.R. Environmental contour lines: A method for estimating long term extremes by a short term analysis. *Trans. Soc. Naval Archit. Mar. Eng.* **2009**, *116*, 116–127.
21. Rosenblatt, M. Remarks on a Multivariate Transformation. *Ann. Math. Stat.* **1952**, *23*, 470–472. [CrossRef]
22. Timco, G.; Weeks, W. A review of the engineering properties of sea ice. *Cold Reg. Sci. Technol.* **2010**, *60*, 107–129. [CrossRef]
23. Shamaei, F.; Bergstrøm, M.; Li, F.; Taylor, R.; Kujula, P. Local pressures for ships in ice: Probabilistic analysis of full-scale line load data. *Mar. Struct.* **2020**, *74*, 102822. [CrossRef]
24. Strub-Klein, L.; Sudom, D. A comprehensive analysis of the morphology of first-year sea ice ridges. *Cold Reg. Sci. Technol.* **2012**, *82*, 94–109. [CrossRef]
25. Ni, B.-Y.; Chen, Z.-W.; Zhong, K.; Li, X.-A.; Xue, Y.-Z. Numerical Simulation of a Polar Ship Moving in Level Ice Based on a One-Way Coupling Method. *J. Mar. Sci. Eng.* **2020**, *8*, 692. [CrossRef]
26. Huang, L.; Tuhkuri, J.; Igrec, B.; Li, M.; Stagonas, D.; Toffoli, A.; Cardiff, P.; Thomas, G. Ship resistance when operating in floating ice floes: A combined CFD&DEM approach. *Mar. Struct.* **2020**, *74*, 102817. [CrossRef]
27. Su, B.; Riska, K.; Moan, T. A numerical method for the prediction of ship performance in level ice. *Cold Reg. Sci. Technol.* **2010**, *60*, 177–188. [CrossRef]
28. Tan, X.; Su, B.; Riska, K.; Moan, T. A six-degrees-of-freedom numerical model for level ice–ship interaction. *Cold Reg. Sci. Technol.* **2013**, *92*, 1–16. [CrossRef]
29. Huisman, M.; Janßen, C.F.; Rung, T.; Ehlers, S. Numerical simulation of ship-ice interactions with physics engines under consideration of ice breaking. In Proceedings of the 26th International Ocean and Polar Engineering Conference, Rhodes, Greece, 26 June–1 July 2016.

30. Lu, W.; Lubbad, R.; Løset, S. Simulating Ice-Sloping Structure Interactions with the Cohesive Element Method. *J. Offshore Mech. Arct. Eng.* **2014**, *136*, 031501. [CrossRef]
31. Ranta, J.; Polojärvi, A.; Tuhkuri, J. Ice loads on inclined marine structures—Virtual experiments on ice failure process evolution. *Mar. Struct.* **2018**, *57*, 72–86. [CrossRef]
32. Gong, H. Discrete-Element Modelling of Ship Interaction with Unconsolidated Ice Ridges: Ridge Resistance and Failure Behaviour. Ph.D. Thesis, Aalto University, Espoo, Finland, 2021.
33. Sand, B.; Horrigmoe, G. Simulations of ice ridge forces on conical structures. In Proceedings of the 8th International Offshore and Polar Engineering Conference, Montreal, QC, Canada, 24–29 May 1998.
34. Myland, D. Ships Breaking Through Sea Ice Ridges. *Int. J. Offshore Polar Eng.* **2014**, *24*, 28–34.
35. Dalane, O.; Aksnes, V.; Løset, S. A Moored Arctic Floater in First-Year Sea Ice Ridges. *J. Offshore Mech. Arct. Eng.* **2015**, *137*, 011501. [CrossRef]
36. Kujala, P. *On the statistics of Ice Loads on Ship Hull in the Baltic*; Helsinki University of Technology: Espoo, Finland, 1994.
37. Lemee, E.; Brown, T. Review of ridge failure against the confederation bridge. *Cold Reg. Sci. Technol.* **2005**, *42*, 1–15. [CrossRef]
38. Kuuliala, L.; Kujala, P.; Suominen, M.; Montewka, J. Estimating operability of ships in ridged ice fields. *Cold Reg. Sci. Technol.* **2017**, *135*, 51–61. [CrossRef]
39. Chai, W.; Leira, B.J.; Næss, A.; Høyland, K.V.; Ehlers, S. Development of Environmental Contours for Ice Ridge Statistics Applied to Reliability-Based Design of Arctic Ships. *Struct. Saf.* **2020**, *87*, 1019906. [CrossRef]
40. Timco, G.; O'Brien, S. Flexural strength equation for sea ice. *Cold Reg. Sci. Technol.* **1994**, *22*, 285–298. [CrossRef]
41. Bonath, V.; Patil, A.; Fransson, L.; Sand, B. Laboratory testing of compressive and tensile strength on level ice and ridged ice from Svalbard region. In Proceedings of the International Conference on Port and Ocean Engineering under Arctic Conditions, Espoo, Finland, 9–13 June 2013.
42. Strub-Klein, L. A Statistical Analysis of First-Year Level Ice Uniaxial Compressive Strength in the Svalbard Area. *J. Offshore Mech. Arct. Eng.* **2017**, *139*, 011503. [CrossRef]
43. Samardžija, I.; Høyland, K.V. Analysis of relationship between level ice draft, ridge frequency and ridge keel draft for use in probabilistic assessment of ice ridge loads on offshore structures. *Cold Reg. Sci. Technol.* **2020**, submitted.
44. Wadhams, P. A comparison of sonar and laser profiles along corresponding tracks in the Arctic Ocean. In *Sea Ice Process and Models, Proceedings of the Arctic Ice Dynamics Joint Experiment International Commission on Snow and Ice Symposium*; Pritchard, R.S., Ed.; University of Washington Press: Seattle, WA, USA, 1980; pp. 283–299.
45. Ervik, Å.; Høyland, K.V.; Shestov, A.; Nord, T.S. On the decay of first-year ice ridges: Measurements and evolution of rubble macroporosity, ridge drilling resistance and consolidated layer strength. *Cold Reg. Sci. Technol.* **2018**, *151*, 196–207. [CrossRef]
46. Wang, Q.; Li, Z.; Lei, R.; Lu, P.; Han, H. Estimation of the uniaxial compressive strength of Arctic sea ice during melt season. *Cold Reg. Sci. Technol.* **2018**, *151*, 9–18. [CrossRef]
47. Arpiainen, M.; Kiili, R. *Arctic Shuttle Container Link from Alaska US to Europe*; Aker Arctic Technology: Helsinki, Finland, 2006.
48. Liu, P.-L.; Der Kiureghian, A. Multivariate distribution models with prescribed marginals and covariances. *Probabilistic Eng. Mech.* **1986**, *1*, 105–112. [CrossRef]
49. Haugen, I. Analysis of Hull Structure Response for a Sevan Arctic Mobile Drilling Unit in Arctic Areas Subjected to Ice Loading. Master's Thesis, Norwegian University of Science and Technology, Trondheim, Norway, 2014.
50. Raghuvanshi, U.; Ehlers, S. An Assessment of Iceberg Loads for Operations in the North Barents Sea. In *Proceedings of the Volume 9: Ocean Renewable Energy*; ASME International: Washington, DC, USA, 2015.
51. Fuglem, M.; Jordaan, I.; Crocker, G.; Cammaert, G.; Berry, B. Environmental factors in iceberg collision risks for floating systems. *Cold Reg. Sci. Technol.* **1996**, *24*, 251–261. [CrossRef]
52. Løset, S.; Onshus, D. Analysis of speeds of drift ice in the Pechora Sea. In Proceedings of the 4th International, Development of the Russian Arctic Offshore (RAO '99), Saint Petersburg, Russia, 6–9 July 1999; Volume I, pp. 248–253.
53. Gagnon, R. Analysis of data from bergy bit impacts using a novel hull-mounted external Impact Panel. *Cold Reg. Sci. Technol.* **2008**, *52*, 50–66. [CrossRef]
54. Pedersen, P.T.; Zhang, S. On impact mechanics in ship collisions. *Mar. Struct.* **1998**, *11*, 429–449. [CrossRef]
55. Ståhlberg, K.; Goerlandt, F.; Ehlers, S.; Kujala, P. Impact scenario models for probabilistic risk-based design for ship–ship collision. *Mar. Struct.* **2013**, *33*, 238–264. [CrossRef]
56. IMO. International code for ships operating in polar waters (Polar Code). In *MEPC 68/21/Add.1 Annex 10*; International Maritime Organization: London, UK, 2015.
57. Bergström, M. A Simulation-Based Design Method for Arctic Maritime Transport Systems. Ph.D. Thesis, Norwegian University of Science and Technology, Trondheim, Norway, 2017.
58. Bergström, M.; Idrissova, S.; Farhang, S.; Huuhtanen, J.; Li, F.; Hirdaris, S.; Ni, B.; Kujala, P. Some new insights towards goal-based design of Arctic ships. In Proceedings of the 8th Transport Research Arena TRA 2020, Helsinki, Finland, 27–30 April 2020.
59. Foo-Nielsen, A.C. *Guidelines for Assessing Ice Operational Risk*; Transport Canada: Ottawa, ON, Canada, 2019.
60. Transport Canada. *Arctic Ice Regime Shipping System Standard TP 12259 (AIRSS)*; Transport Canada: Ottawa, ON, Canada, 2018.

applied sciences

MDPI

Article

Simulation-Based Assessment of the Operational Performance of the Finnish–Swedish Winter Navigation System

Martin Bergström * and **Pentti Kujala**

Department of Mechanical Engineering, Aalto University, FI-00076 Aalto, Finland; pentti.kujala@aalto.fi
* Correspondence: martin.bergstrom@aalto.fi; Tel.: +358-50-476-7229

Received: 15 August 2020; Accepted: 24 September 2020; Published: 27 September 2020

Abstract: This article presents a discrete event simulation-based approach for assessing the operating performance of the Finnish–Swedish Winter Navigation System (FSWNS) under different operating scenarios. Different operating scenarios are specified in terms of ice conditions, the volume of maritime traffic, number of icebreakers (IBs), and regulations such as the Energy Efficiency Design Index (EEDI). Considered performance indicators include transport capacity, number of instances of icebreaker (IB) assistance, and IB waiting times. The approach is validated against real-world data on maritime traffic in the Bothnian Bay. In terms of the number of ship arrivals per port, indicating the transport capacity of the FSWNS, the simulation agrees well with the data. In terms of the number of instances of IB assistance and IB waiting times per port, the standard deviations between the mean of 35 independent simulation runs and the data are 13% and 18%, respectively. A sensitivity analysis indicates that the simulated number of instances of IB assistance and IB waiting times is particularly sensitive to assumptions concerning the presence of brash ice channels. Case studies indicate that, unless the number of IBs is increased, the EEDI regulations may result in a significant increase in both the number of instances of IB assistance and the cumulated IB waiting times.

Keywords: winter navigation; arctic shipping; ice class ships; icebreakers; maritime transport systems; energy efficiency design index (EEDI); discrete-event simulation

1. Introduction

The Baltic Sea is an important transit route connecting numerous countries and markets. In 2019, the total volume of Finnish import and export transported over the Baltic Sea exceeded 100 million tons, corresponding to around 80% of the total trade [1,2]. These numbers are expected to increase as the long-term trend in the volume of Finnish seaborne trade is one of growth [3].

In winter, large parts of the Baltic Sea are typically ice-covered, but with significant interannual variability with regards to the maximum ice extent [4]. This variability poses a challenge to shipping in the region, as sea ice has a significant impact on the operations and transit times of ships. The aim of the Finnish–Swedish Winter Navigation System (FSWNS) is to maintain safe and efficient year-round navigation to and from Finnish and Swedish ports along the Baltic Sea [5]. To this end, the FSWNS manages winter navigation-related challenges by the combined use of (a) ice class rules, (b) traffic regulations, and (c) icebreaker (IB) assistance [6]. Specifically, to make sure that ships have enough ice-going capability for safe and efficient operations, they must be built and operated following the Finnish–Swedish Ice Class Rules (FSICR) [5]. These are enforced by port-specific traffic restrictions set by Finnish and Swedish maritime authorities in terms of the minimum ice class and deadweight needed to be eligible for IB assistance [5]. IB assistance is provided based on the available fleet of Finnish and Swedish state-owned and operated IBs. As per [7,8], Finland has a fleet of eight major IBs

(Polaris, Fennica, Nordica, Otso, Kontio, Voima, Sisu, and Urho), whereas Sweden has a fleet of five major IBs (Ale, Atle, Frej, Oden, and Ymer).

Both in the short- and longer-term, the performance of the FSWNS is expected to be influenced by the International Maritime Organization's (IMO) Energy Efficiency Design Index (EEDI) regulations. These regulations, which were adopted by the IMO in July 2011, aim to reduce the amount of greenhouse gasses (GHGs) from ships by promoting the use of more energy-efficient solutions [9]. Anyhow, due to the technical content of the regulations, they are expected to limit the installed propulsion power of ships, which will reduce their ice-going capability and attainable speed in ice. This in turn could increase the demand for IB assistance, resulting in more frequent and longer waiting times for IB assistance, which is expected to increase the overall transport costs and time.

Towards assessing the influence of the EEDI, and to identify and assess possible mitigation measures, this article presents an approach for assessing the operational performance of the FSWNS under various operating condition scenarios. The approach builds on a discrete event simulation (DES)-based approach presented by [10]. DES is a specific type of simulation in which the behavior of a system is modelled as an ordered sequence of events, each of which takes place at a specific point of time and results in a change in the state of the system [11]. As no change occurs between events, DES enables fast simulations of extensive operating periods. Additionally, as individual events can be modelled so that they are dependent on stochastic variables, DES is well suited to considering uncertainties and stochastic factors. In addition, because DES makes it possible to model individual ships as entities moving through a system in the manner of a queue, the technique is well suited to capturing various interactions and self-reinforcing effects.

The approach presented by [10] can assess the performance of an Arctic maritime transport system consisting of a homogenous fleet of ships operating between two ports. In this study, the approach is extended to handle the significantly more complex operations of the FSWNS. Related research questions include the following: Is it feasible to capture the complex behavior and to roughly estimate the operating performance of the FSWNS using a DES-based approach? What behaviors of the FSWNS are difficult to capture using DES? What are the future research needs?

Other approaches to simulate the FSWNS include [12], in which the transport system is simulated using a brute-force optimization-based approach, providing a good agreement with real-world data. A potential weakness of [12] is that its application appears to be laborious and time-consuming due to its mathematical nature. Furthermore, as the approach [12] appears not to be based on ship entities, it is unclear how it handles various interaction and self-reinforcing behaviors of the FSWNS (e.g., if a ship arrives late at a port, it should also leave the port late), which might be significant in an extreme scenario.

2. Background

2.1. Finnish–Swedish Ice Class Rules and IB Waiting Time

The aim of the Finnish–Swedish Ice Class Rules (FSICR) is to ensure that ships operating on the northern Baltic Sea, to and from Finnish and Swedish ports, have sufficient ice-going capability to maintain safe and efficient navigation year-round [13]. To this end, the rules, which have been developed jointly by the Finnish and Swedish maritime authorities based on accumulated experience and research, specify five ice classes: IA Super, IA, IB, IC, and II. Enforcement is through port-specific restrictions determining the minimum ice class and deadweight needed to be eligible for IB assistance [13].

The demand for IB assistance in a region depends among others on the prevailing ice conditions, the amount of maritime traffic, and the ice-going capability of the ships operating there. If the demand for IB assistance exceeds the available icebreaking resources, the waiting time for IB assistance will increase. To maintain smooth and efficient maritime traffic, the goal of the Finnish IB service is to limit the average waiting time to four hours [14]. To this end, the FSICR determines ice class-specific performance requirements. These are determined as per [5] in terms of the minimum ice conditions in which a ship must be able to maintain a speed of at least 5 knots.

2.2. Energy Efficiency Design Index (EEDI)

The EEDI regulations regulate a ship's CO_2 emissions by specifying its maximum allowed EEDI value determined as a function of deadweight tonnage (DWT) or gross tonnage (GT), separately for different types of ships (e.g., bulk carriers, tankers, gas carriers, roll-on/roll-off cargo ships) [15]. To stimulate continued innovation and technical development, the maximum allowed EEDI will be tightened incrementally every five years [9].

In simplified terms, the EEDI value represents the amount of CO_2 generated by a ship carrying out a specific transport work [15]. Accordingly, the EEDI value can be expressed as per Equation (1) based on engine power, specific fuel consumption (SFC), an assumed amount of CO_2 per gram of fuel (C_F), DWT, and ship speed [16].

$$\text{EEDI} = \frac{\text{CO}_2\text{ emissions}}{\text{transport work}} = \frac{\text{Engine power} * C_F}{\text{DWT} * \text{speed}} \tag{1}$$

As per Equation (1), for a given type of engine and fuel, the EEDI regulations effectively limit the maximum installed propulsion power. To make ice-class ships comparable with open water ships, considering that they need extra propulsion power for operation in ice, the EEDI regulations include correction factors [17]. These are determined for five different types of ships: tanker, bulk, general cargo, container, gas carriers, and roll-on/roll-off ships [17]. Notwithstanding, the EEDI regulations are expected to reduce the average propulsion power, and consequently also the average ice-going capability of ice classed ships. This implies a reduction in both the maximum ice conditions in which ships can operate independently and the speed of ships in ice. For a given operating scenario, this will increase both the number of instances where a ship needs IB assistance and the duration of each instance of IB assistance. As a result, the demand for IB assistance is expected to increase [18]. Additionally, considering the maritime industry's overall efforts to optimize maritime operations, the demand for IB assistance might also be driven by non-EEDI-related cost and energy consumption reducing measures, further reducing the ice-going capability of ships.

3. Description of the Simulation Model

3.1. Programming Platform

The proposed simulation approach is implemented in MATLAB (ver. R2020a), using its discrete event simulation tool SimEvents (based on SimuLink ver. 10.1/ R2020a). As per [19], SimEvents provides a discrete-event simulation engine and component library for analyzing event-driven system models and optimizing performance characteristics. A set of predefined blocks, including queues, servers, switches, supports the modelling work.

3.2. Model Structure and Working Principle

As per Figure 1 presenting an example simulation model structure, the simulation model consists of different types of blocks representing navigation legs (L), ports, crossings, borders between different IB operating areas, and ship entry/exit gates. Ships are represented by entities, each of which has a set of predetermined attributes specifying the technical characteristics (e.g., ice-going capability) and voyage characteristics (destination ports, port-turnaround times) of the ship. IBs, on the other hand, are represented by "resources". Specifically, an individual IB is represented by an individual IB resource that, when assisting a ship, is attached to the ship being assisted.

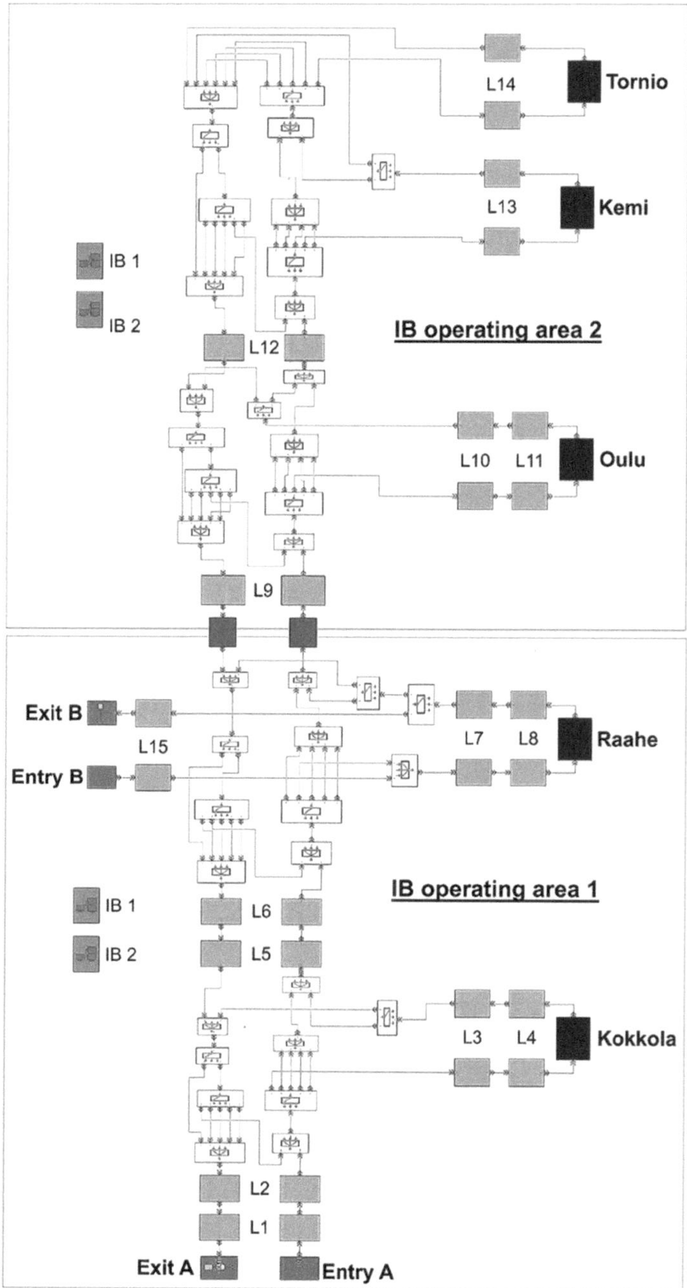

Figure 1. Simulation model example. Entry/Exit A is used by ships arriving from or leaving towards Kvarken, Entry/Exit B is used by ships arriving from or leaving towards Lulea (Sweden). Letter L stands for (navigation) leg.

Each ship entity enters the simulation model at a specific date and time (determined based on maritime traffic data) through an entry gate with an assumed geographical location (e.g., Kvarken, Bay of Bothnia). Once entered, a ship entity will progress towards its first port of destination, where it will stay for a predetermined period corresponding to the total port turnaround time. Thereafter, the ship entity will either continue towards another port within the simulation model or towards a port located outside the simulation model. In the latter case, the ship entity will progress towards an exit gate with an assumed geographical location and then leave the simulation model.

Navigation legs are here defined as the geographical distance between two waypoints. The time it takes for a ship entity to complete a leg depends on the leg distance, the ice conditions, the operating mode (independent or assisted operation), the ship's estimated speed in the prevailing ice conditions and operating mode, and the waiting time for IB assistance (in case the ship must call for IB assistance). Specifically, navigation legs are modelled as per the schematic diagram in Figure 2 whose various elements are described as follows:

- A—Date definition. When a ship entity (with or without IB assistance) arrives at a waypoint, the present date is determined in terms of the number of days elapsed since the start of the simulation.
- B—Ice conditions. The prevailing ice conditions are determined following a predefined table defining the ice conditions by navigation leg and date. The prevailing ice thickness along the leg is defined in terms of the average equivalent ice thickness (H_{eq_avg}) (cm) and the maximum equivalent ice thickness (H_{eq_max}) (cm). H_{eq_avg} is defined as the average thickness of all major ice features (level ice, ice ridges, openings) over the whole leg. H_{eq_max}, in turn, is defined as the average thickness of the same ice features over the part of the leg with the most difficult ice conditions (e.g., an area with severe ice ridging). In order to account for uncertainty and stochasticity, during an individual simulation run the applied H_{eq_avg} and H_{eq_max} values are multiplied by randomly determined coefficients representing their uncertainty. In addition, based on the location and prevailing ice conditions, an assumption is made as to whether a brash ice channel is present. Specifically, depending on the location of a leg and the prevailing ice conditions there, a brash ice channel is assumed to be (a) present at all times, (b) present with a certain probability, or (c) never present. If an ice channel is assumed present with a certain probability, whether an ice channel is present at a specific date is determined based on a binary number (0 = no ice channel, 1 = ice channel) drawn from an assumed distribution.
- C—Speed without IB assistance. The assumed independently achievable speed (knots) of a vessel is determined both for H_{eq_avg} and H_{eq_max} based on ship and operation type-specific hv curves that determine the speed of a ship as a function of the ice thickness. As per the example hv curves presented in Figure 3, two different types of independent operation are considered:
    - Independent operation in a brash ice channel ("Channel" as per Figure 3). Here, the ship is operating in a pre-existing brash ice channel without IB assistance. Ice resistance is higher than when operating with IB assistance because broken ice is distributed over the channel area. H_{eq} relates to the prevailing thickness of the unbroken ice in the area.
    - Independent operation in level ice or through a large ice floe ("Level ice" as per Figure 3).
- D—Need for IB assistance. Whether a ship needs IB assistance (or continued assistance in case the ship is already assisted by an IB) to complete the next upcoming leg is determined based on its estimated independently achievable speed (knots) in the worst expected ice conditions (H_{eq_max}) along the leg (calculated in block C). If a ship is not assisted by an IB, the ship will stop and call for IB assistance if its estimated independently achievable speed in H_{eq_max} falls below a defined threshold (e.g., 1.5 knots). Otherwise, the ship is considered able to continue independently. If a ship is assisted by an IB from before, the assistance will continue until the ship's independently achievable speed in H_{eq_max} exceeds another higher threshold value (e.g., 8 knots). This means that an IB is assumed not to leave an assisted ship in ice conditions in which it can barely continue

independently, but to assist a ship until it has reached open water or ice conditions in which it can continue independently without difficulty.

- E—Junction 1. A ship entity's choice of path at Junction 1 depends on whether the ship that it represents is considered to be in need of IB assistance. If the represented ship is considered able to continue independently, the ship entity continues to block F. In this case, if the ship is assisted by an IB, the resource representing the assisting IB is released from the ship and becomes available to assist other ships. On the other hand, if the represented ship is considered to require IB assistance, the ship will proceed to block G.
- F—Leg time without IB assistance. In the case of independent operation, the time a ship needs to complete a leg is calculated in hours based on the leg distance and the ship's independently achievable speed (determined in block C). The ship entity will remain in the block for a period corresponding to the calculated leg time.
- G—Junction 2. A ship entity's choice of path at junction 2 depends on whether the ship that it represents is assisted by an IB. If the ship is assisted by an IB, the ship entity continues to block L. Otherwise, it continues to block H.
- H—Convoy formation. Typically, IBs assist ships one by one. However, if multiple ships need IB assistance over the same distance at the same time, a convoy operation can be carried out in which an IB assists more than once ship at a time. Convoy operations are challenging as they, among others, require a significant safety distance between the assisted ships [20]. This limits the feasible convoy length as the hull ice resistance of assisted ships tends to increase as a function of the distance to the escorting IB. In the present simulation model, an IB is assumed to assist a maximum of two ships at a time. Specifically, a convoy is formed if a ship entity arrives at block H while another ship entity is already waiting for IB assistance in block I. In that case, the arriving ship entity proceeds to block K. If another ship entity is already waiting in block K, the ship proceeds to block I.
- I—Acquisition of IB assistance. A ship entity arriving at block I will trigger a call for IB assistance and wait until an IB resource becomes available. Once a ship entity has been assigned an IB it will proceed to block J. In case multiple ships are waiting for assistance in block I, the ships will be assisted in the order in which they arrived.
- J—IB transfer and maneuvering time. Although an IB is assumed to remain within its operating area at all times, the exact position from which an IB starts to move towards a ship calling for assistance is not known. Therefore, the related "transfer time" (hours) is determined probabilistically based on an assumed distribution. Once an IB has reached a ship or convoy in need of assistance, before the assistance may start, the IB must maneuver itself into an appropriate position ahead of the ship(s) that are to be assisted. Additionally, in case the ship(s) are stuck in ice, the icebreaker must first cut loose the ship(s) by breaking the surrounding ice. The corresponding maneuvering time is determined probabilistically based on an assumed distribution. The ship entity will remain in the block for a time corresponding to the sum of the determined IB transfer and maneuvering times.
- K—Ship waiting to join a convoy. A ship entity waiting in block K will form a convoy with the first IB-assisted ship entity exiting block J. The IB-assisted ship entity arriving from block J is assumed to have priority access to the IB, meaning that it will be assisted by the IB until it has reached its destination or is deemed able to continue independently (see description of block D—"Need for IB assistance"). The ship entity joining the convoy from block K, on the other hand, will be assisted over one leg only after which it needs to call for further IB assistance if needed.
- L—Speed with IB assistance. The speed (knots) of a ship assisted by an IB is determined as the lower of the achievable speed of the assisted ship and the achievable speed of the assisting IB. In other words, the speed is either limited by the assisted ship or by the IB. The achievable speed of the assisted ship is determined based on ship model-specific hv curves for "Assistance at distance", examples of which are presented in Figure 3. The achievable speed of the IB is determined as per

the description of block C—"Speed without IB assistance". In the case of convoy operation, as a simplification, the speed of each assisted ship is determined separately as described above.

- M—Leg time with IB assistance. The leg time (hours) for an IB-assisted ship or convoy is determined based on the leg distance and the speed as determined in block L. The ship entity or entities will remain in the block for a time corresponding to the calculated leg time.

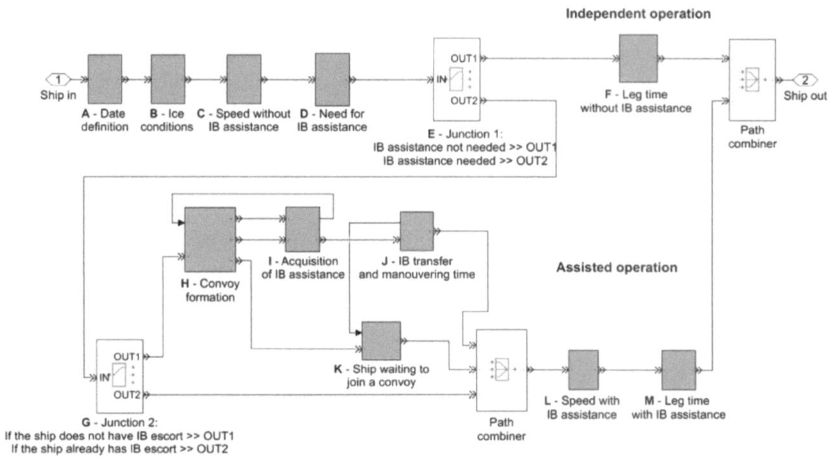

Figure 2. Leg model structure.

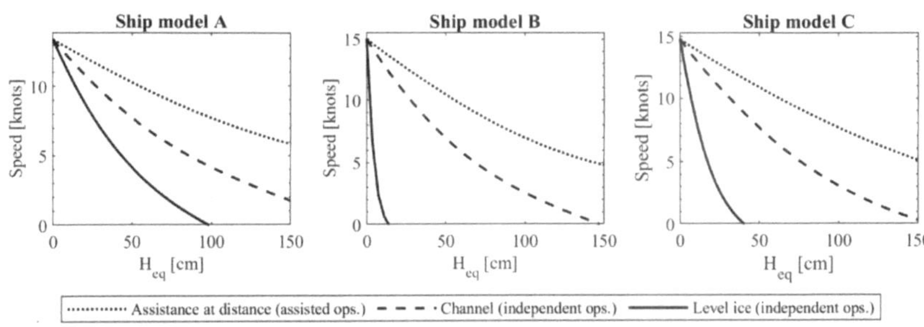

Figure 3. Examples of hv curves for different ship models. The hv curves determine the speed of a ship as a function of the equivalent ice thickness (H_{eq}).

The performance of the FSWNS depends on many stochastic factors. In the simulation, as explained above, random variables are generated by drawing random numbers from variable-specific distributions. It can be noted that using MATLAB, the generation of streams of random values is as default repeatable, meaning that separate simulation runs result in identical streams of random values. This is well suited to analyze the influence of variations in single parameter values or distributions. However, in order to analyze how the system performs under different combinations of parameter values, independent sets of random values must be generated.

A general challenge related to the modelling of stochastic factors of the FSWNS is the lack of relevant data. In this study, four stochastic variables are considered: (a) the presence of brash ice channels as determined in block B, (b) the time it takes for an IB to reach a ship in need of assistance as determined in block J, (c) the duration of the IB maneuverings required before an IB assistance can start, and (d) uncertainty in the assumed H_{eq_avg} and H_{eq_max} values. Examples of how these variables can be specified are presented in Section 4.

Using DES, the simulated period is divided into equal time steps. In this study, the length of the applied time step is one hour. This means that a change in the state of the system is occurring every hour. Another important time unit is the number of days since the start of the simulation, based on which date is defined. The date is important as it, as explained above, defines the prevailing ice conditions along a specific leg and whether a brash ice channel is present.

As per Figure 1, IBs are assumed to operate within a limited IB operating area. In the simulation, this means that once an IB resource has assisted a ship entity to the border of its operating area, it will leave the ship. If further assistance is needed, a ship entity will request an IB resource from within the IB operating area that it is entering. Thus, in line with the available maritime traffic data, a ship entity might be assisted by several different IB resources on its way towards its destination. In this case, the total waiting time for IB assistance is the accumulated sum of the waiting times related to each instance of IB assistance.

3.3. Generalizations and Assumptions

Resulting from the complexity of the FSWNS, as well as due to various identified knowledge gaps and technical limitations of the applied simulation technique, the simulation model simplifies and generalizes some of the characteristics and mechanisms of the FSWNS. Specifically, the following generalizations and assumptions are noted:

- Concept of equivalent ice thickness. As applied in this approach, the concept of equivalent ice thickness rests on the assumption that an ice cover of a specific equivalent thickness results in the same level of hull resistance as continuous level ice of the same thickness [21]. A weakness of this concept is that it fails to account for individual ice features (e.g., individual ice ridges) that might stop a ship [22]. Anyhow, based on [23], for the simulation of ships operating on the Baltic Sea, the concept appears well suited.
- The use of hv-curves. The use of hv curves to model the speed of ships in ice is well established. It should be noted that this approach typically rests on the assumption that ships operate at a fixed engine load (typically near the maximum continuous rating), which is not always true [24]. However, currently, there is no general and publicly available approach to eliminate this assumption.
- Multi-ship convoy operations. In the simulation, convoy operations in which an IB assists two ships at a time occur whenever two ships need IB assistance over the same distance at the same time. Convoy operations in which three or more ships are assisted at one time are not considered. These simplified assumptions are needed as it is not entirely clear under what conditions convoy operations may occur. The ice resistance of a ship being assisted by an IB tends to increase as a function of the distance between the ship and the assisting IB. Therefore, because a significant safety distance is required between ships operating in a convoy, multi-ship convoy operations require a higher ice-going capability from the involved ships [20]. As a result, considering the EEDI regulations and other measures lowering the average ice-going capability of modern ships, convoy operations and particularly those in which an IB assists more than two ships at a time are expected to remain rare in the future, especially during periods of heavy ice conditions. For this reason, the exclusion of convoy operations involving more than two assisted ships is not expected to significantly reduce the accuracy of the simulation model when applied to simulate heavy ice condition scenarios.
- IB transit times. Due to limitations set by the applied simulation technique, the exact location from which an available IB starts to move towards a ship in need of IB assistance is not known. Therefore, the duration of the IB transit is determined probabilistically based on statistics. In the real world, the master of an IB may try to minimize a ship's waiting time by predicting where assistance will be needed, and if possible, start to proceed towards that area in advance. Thus, particularly during periods of low demand for IB assistance, the above-described approach is likely conservative. On the other hand, in periods of high demand for IB assistance, IB waiting

times appear to be primarily driven by the availability of IBs, meaning that the relationship between transit times and ships' total IB waiting times is small.

- Criteria for providing of IB assistance. In the simulation, IB assistance is provided if a ship's independently achievable speed falls below a specific limit value (e.g., 1.5 knots) in the worst assumed ice conditions along a leg. In the real-world, the criteria for IB assistance likely depends on the operating situation so that the criteria are stricter during times of high demand for IB assistance. In addition, the decision on whether IB assistance is to be provided, or requested, is also likely influenced by the individual judgement of the masters of the involved ships. Currently, there are no publicly available models or principles based on which such decision-making could be modelled accurately.
- Active measures by the crew. As per the simulation model structure presented in Figure 1, the network of routes along which ships operate throughout a simulation is assumed fixed. This means that active crew measures, such as maneuverings to avoid local areas with difficult ice conditions, are not considered. As a result, particularly for sea areas with partially ice-covered waters, the simulation outcome can be assumed conservative. In principle, this limitation could be overcome, e.g., by applying a voyage optimization tool as proposed by [25]. However, this would make the approach significantly more complex.

4. Validation

4.1. Approach

Validation of the model is carried out based on real-world maritime traffic data obtained through the research project WINMOS II (Winter Navigation Motorways of the Sea II) [26] covering maritime traffic on the Bothnian Bay in the period 15 January–5 February 2010. The accuracy of the model is assessed by comparing simulated and data-based performance indicators, such as the number of port arrivals, the number of instances of IB assistance, and IB waiting times. To capture the stochastic behavior of the system, a total of 35 individual simulation runs are carried out. All considered maritime traffic data are presented in the Appendix A (see Tables A1 and A2).

4.2. Simulation Input

4.2.1. Maritime Traffic

Information on ships entering the Bothnian Bay is specified based on the above-mentioned maritime traffic data, an extract of which is presented in Table 1. The entry time is specified in hours from the start of the simulation at midnight 15 January 2010. As per Figure 4, ships arriving from the South (Kvarken) are assumed to enter the considered area at point A, whereas ships arriving from the Northwest (northern Sweden, Lulea) are assumed to enter at point B. Entered ships visit 1–3 ports before they leave the system. The total duration of each port visit is determined in terms of the port turnaround time (PTT) as specified by the maritime traffic data.

Table 1. Extract of maritime traffic data. The port turnaround time (PTT) represents the total duration of a port visit including port maneuvering time, mooring time, and cargo handling time.

Entry Time, Point A (h)	Ship Model	Port 1	Port 2	Port 3	PTT 1 (h)	PTT 2 (h)	PTT 3 (h)
76	F	Kokkola			73		
83	G	Raahe	Tornio		64	19	
91	C	Oulu			19		
91	G	Tornio	Kotka		25		
91	G	Kemi			28		
92	G	Kokkola	Kemi		60	53	

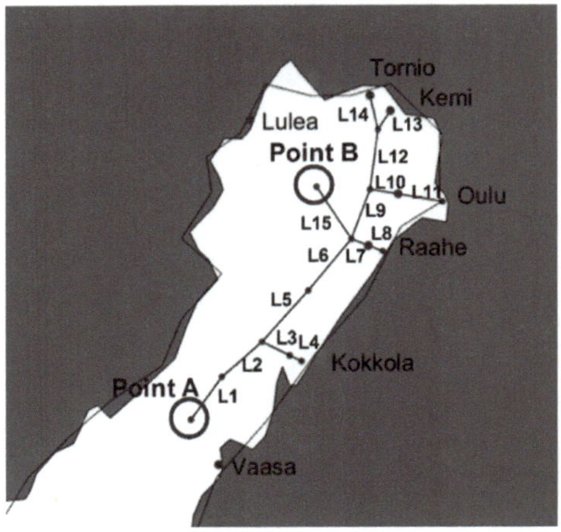

Figure 4. Navigation legs (L) and ports.

4.2.2. Ship Performance Data

The considered maritime traffic data covers ships representing some 16 different models. For each ship model, as per Figure 5, the achievable speed in ice is determined in terms of ship model-specific hv curves representing three different modes of operation: (a) IB assistance at distance, (b) operation in a brash ice channel, and (c) independent operation in level ice. All hv curves were determined within the research project WINMOS II [26] based on ice resistance formulas and design particulars of the corresponding real-world ships.

4.2.3. Ice Conditions

Based on ice charts provided by [27], for each considered date and navigation leg, the prevailing ice conditions are determined in terms of the average ice conditions H_{eq_avg} and the most difficult ice conditions H_{eq_max}. Specifically, the average ice conditions along a leg H_{eq_avg} is determined as per Equation (2).

$$H_{eq\_avg} = \frac{H_{min} + H_{max}}{2} * \frac{C_{min} + C_{max}}{2} + H_{ridging\_avg}, \tag{2}$$

where H_{min} represents the minimum ice thickness, H_{max} represents the maximum ice thickness, C_{min} represents the minimum ice concentration, C_{max} represents the maximum ice concentration, and $H_{ridging\_avg}$ is an assumed 6 cm increase in equivalent ice thickness caused by ridges (where ice ridging occurs). The assumed value of $H_{ridging\_avg}$ is based on [28], according to which ice ridges in the Bothnian Bay increase the level ice thickness by 6–14 cm equivalent ice thickness. The most difficult ice conditions occurring along a leg H_{eq_max} is determined as per Equation (3).

$$H_{eq\_max} = H_{max} * C_{max} + H_{ridging\_max}, \tag{3}$$

where H_{max} represents the maximum ice thickness, C_{max} represents the maximum ice concentration, and $H_{ridging\_max}$ is an assumed 14 cm increase in level ice thickness caused by ridges (where ice ridging occurs) [28].

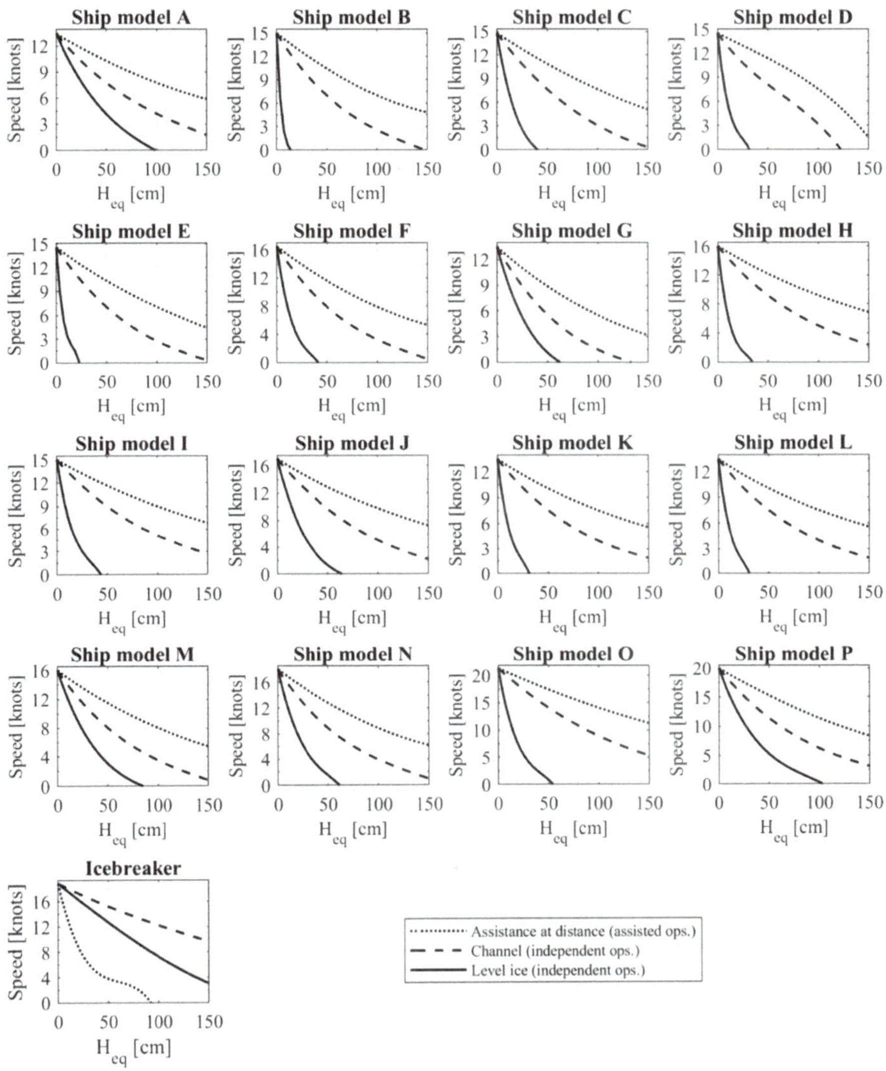

Figure 5. Ship model-specific hv curves for three different operation modes: icebreaker (IB) assistance at distance, operation in a brash ice channel, and independent operation in level ice [26].

Extracts of calculated and applied H_{eq_avg} and H_{eq_max} values are presented in Tables 2 and 3. The complete ice data are found in Appendix B (see Tables A3 and A4). It should be noted that the assumed values of H_{eq_avg} and H_{eq_max} are subject to significant uncertainty originating both from the source, i.e., the ice charts, which presents ice data in terms of approximate thickness and concentration ranges, and from the assumption that the ice conditions are homogenous over the considered navigation legs. To account for this uncertainty, during an individual simulation run, each value of H_{eq_avg} and H_{eq_max} presented in Tables 2 and 3 is multiplied with two coefficients, one representing the uncertainty in ice thickness, and the other representing the uncertainty in ice concentration. Both coefficients are determined by drawing a random number from a normally distributed set with a mean of 1 and a standard deviation of 10%.

Table 2. Extract of H_{eq_avg} (cm) values.

Date	L1	L2	L3	L4	L5	L6	L7	L8	L9	L10	L11	L12	L13	L14	L15
2 February 2010	4	5	5	33	5	5	5	38	21	32	38	21	38	38	43
3 February 2010	4	5	5	33	5	5	5	38	21	32	38	21	38	38	43
4 February 2010	4	5	5	33	5	5	5	38	21	32	38	21	38	38	43
5 February 2010	23	23	23	40	23	23	23	38	23	38	38	39	45	45	45
6 February 2010	23	23	23	40	23	23	23	38	23	38	38	39	45	45	45
7 February 2010	23	23	23	40	23	23	23	38	23	38	38	39	45	45	45

Table 3. Extract of H_{eq_max} values (cm).

Date	L1	L2	L3	L4	L5	L6	L7	L8	L9	L10	L11	L12	L13	L14	L15
2 February 2010	5	5	5	40	5	5	5	50	30	30	50	30	50	50	54
3 February 2010	5	5	5	40	5	5	5	50	30	30	50	30	50	50	54
4 February 2010	5	5	5	40	5	5	5	50	30	30	50	30	50	50	54
5 February 2010	36	36	36	50	36	36	36	50	36	50	50	50	60	60	59
6 February 2010	36	36	36	50	36	36	36	50	36	50	50	50	60	60	59
7 February 2010	36	36	36	50	36	36	36	50	36	50	50	50	60	60	59

Ships operating in ice may encounter brash ice channels created and maintained by IBs and other ships. The duration for which an ice channel remains open and navigable depends on multiple factors including the amount of maritime traffic, the local geography, as well as the prevailing wind and currents [29]. In addition, strong winds in combination with a physical boundary might result in compressive ice, which may significantly increase a ship's ice resistance [30]. Anyhow, due to a lack of related data and suitable engineering models, such factors are not systematically considered. Instead, it is assumed that along fairways with very significant traffic (L 1 and 2), a brash ice channel is available with a 75% probability, along fairways with significant traffic (L 3, 5, 6, 7, 9, 10, and 12), a brash ice channel is available with a 50% probability. On the open sea, where the traffic is limited (L 15), brash ice channels are assumed not to be present. In protected waters with fast ice (L 4, 8, and 11, 13, and 14), brash ice channels are assumed to be present throughout the simulation.

4.2.4. IB Transfer and Maneuvering Times

Due to a lack of available real-world data, the time it takes for an available IB to reach a ship in need of assistance, in the following referred to as IB transfer time, is determined based on statistics of simulated transfer times obtained from research project WINMOS II [26], generated based on the mathematical simulation approach by [12]. Following the obtained data, IB transfer times are assumed to be exponentially distributed as per the probability density function (PDF) in Figure 6. Accordingly, the time it takes for an available IB to reach a ship in need of assistance is assumed to be in the range of 0–8 h. The maximum transfer time may correspond to a situation where an IB must cover some 80 NM at an average speed of 10 knots to reach a ship in need of assistance. This distance is roughly equivalent to, for instance, the distance L1–L4 (see Figure 4). The IB maneuvering time is assumed to follow a normal distribution with a mean value of 20 min and a standard deviation of 5 min.

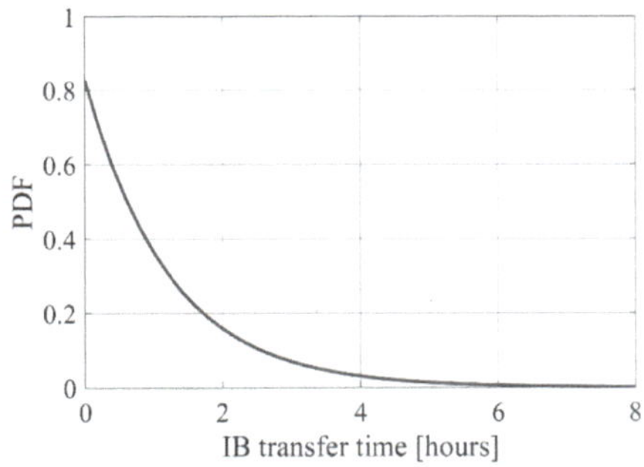

Figure 6. Probability density function (PDF) of IB transfer times determined based on simulated data obtained from Winter Navigation Motorways of the Sea II (WINMOS II) [26].

4.2.5. IB Assistance Parameters

During the considered period 15 January–15 February 2010, as per ice charts issued by [27], IB assistance in the Bothnian Bay is provided by four IBs: Fennica, Urho, Otso, and Kontio. However, as a simplification, all icebreakers are assumed to have an ice-going capability corresponding to that of Otso. The IBs are assumed to operate within their operating area as defined by Figure 1. IB assistance is provided to those ships whose independently achievable speed is estimated to fall below 1.5 knots in the worst ice conditions (H_{eq_max}) along a leg. IB assistance is continued until a ship has been assisted to a waypoint from which it can continue independently at a minimum speed of 8 knots. During icebreaker assistance, the maximum speed of the IB and the assisted ship(s) is assumed limited to 12 knots.

4.3. Validation Results

4.3.1. Real World vs. Simulated Number of Port Visits

Figure 7a presents the real-world number of ship arrivals per destination port, the corresponding simulated values for 35 individual simulation runs, and the mean of the simulated values. As per the figure, the simulated values agree well with the data. This indicates that the traffic flows within the simulation model follow the data and that the simulated ship transit times are in the same range as, or lower than, the real-world transit times.

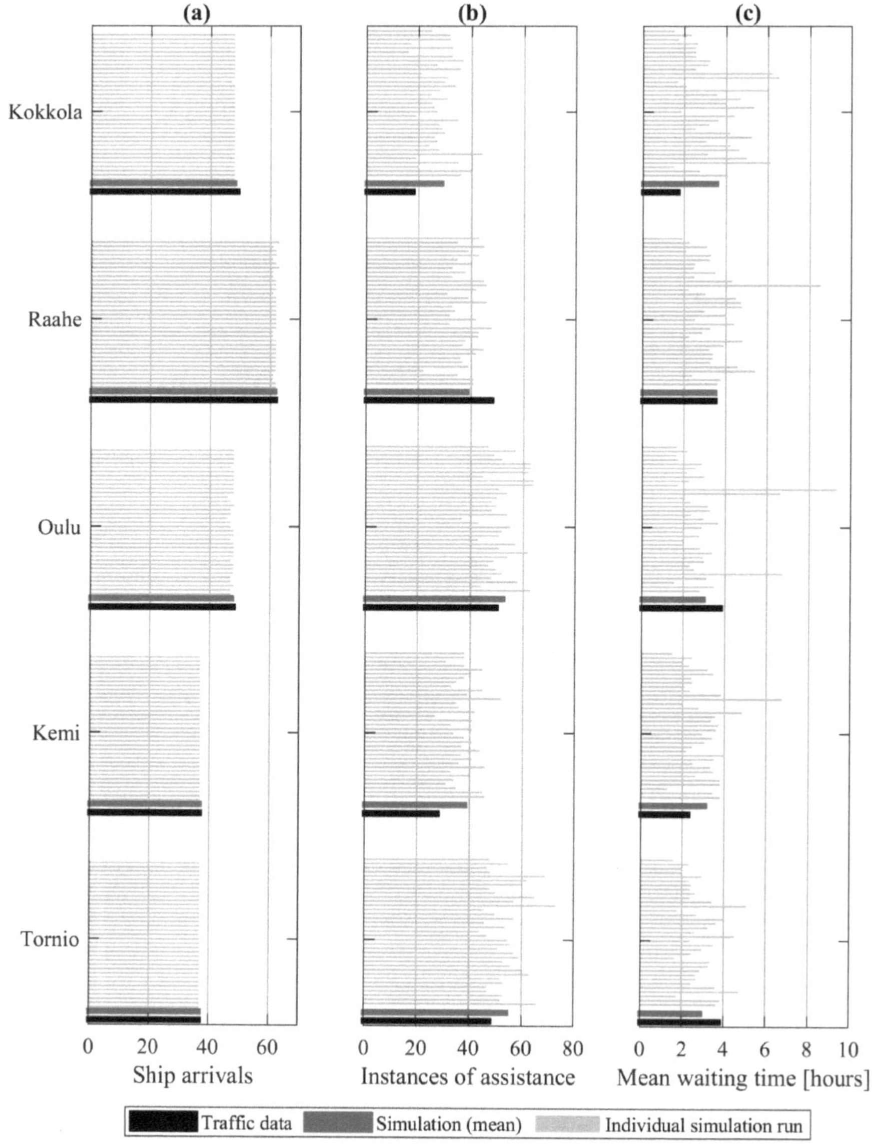

Figure 7. Real-world vs. simulated port-specific number of ship arrivals (**a**), instances of IB assistance (**b**), and mean IB waiting times (**c**). The traffic data indicate the real-world performance of the system during the considered period. The corresponding simulated performance values are presented both in terms of the mean values of 35 independent simulation runs, and in terms of the outcome of each of those individual simulation runs. The outcomes of the individual simulation runs demonstrate the variability of the simulation output.

4.3.2. Real World vs. Simulated Number of Assisted Voyages

Figure 7b presents the real-world number of instances of IB assistance per destination port, the corresponding simulated values for 35 individual simulation runs, and the mean of the simulated values. The presented numbers include all instances of IB assistance received by ships on their way to or from individual ports. As ships are often assisted at multiple times during a single voyage, the number of instances of assistance is higher than the number of ships that received assistance. As shown by the figure, for all ports the mean values of the simulations agree quite well with the data. However, the variation between individual simulation runs is significant. The standard deviation between the mean values and the data is 13%. As per the data, the total number of instances of assistance is 192 whereas the corresponding simulated mean value is 213 (+11%).

4.3.3. Real World vs. Simulated IB Waiting Times

Figure 7c presents the real-world mean IB waiting time per destination port, the corresponding simulated values for 35 individual simulation runs, and the mean of the simulated values. The presented values correspond to the mean waiting time for all instances of IB assistance related to a ship heading to or leaving from a given port. As shown by the figure, in general the data and the mean of the simulated values agree quite well. The deviation is the largest in the case of the port of Kokkola. This is because most ships use the fairway outside Kokkola (Leg 1–2), which in the simulation apparently occasionally results in bottleneck situations. In the real world, such bottleneck situations might be avoided by various factors not considered in the applied simulation model (e.g., in the real world an IB might perhaps if needed assist more than two ships at a time, or the likelihood of an ice channel being present might be higher than assumed during periods of heavy traffic). As for the instances of assistance, the variation between individual simulation runs is significant. The overall standard deviation between the mean values of the individual simulation runs and the data is 18%. As per the data, the mean IB waiting time is 3.0 h, whereas the corresponding simulated number is 3.2 h (+7%).

4.3.4. Confidence Intervals

The statistical agreement between the maritime data and the simulation outcome is tested in terms of confidence intervals. Specifically, for each of the simulated port-specific number of instances of IB assistance and mean IB waiting times, we consider an approximate 95 confidence interval determined by multiplying the standard deviation (σ_{sim}) of the mean values (μ_{sim}) of 35 independent simulation runs by 2. As per Tables 4 and 5, for each of the simulated performance metrics, the maritime data fall within the obtained 95% confidence interval.

Table 4. Approximated 95% confidence interval for the simulated port-specific number of instances of IB assistance.

Port	Maritime Data	μ_{sim}	σ_{sim}	$\mu_{sim} - 2\sigma_{sim}$	$\mu_{sim} + 2\sigma_{sim}$
Kokkola	18	28.7	6.5	15.8	41.7
Raahe	48	38.7	5.5	27.6	49.7
Oulu	50	52.5	6.8	38.8	66.1
Kemi	28	38.4	5.2	28.0	48.9
Tornio	48	54.5	7.1	40.3	68.6

Table 5. Approximated 95% confidence interval for the simulated port-specific mean IB waiting times (hours).

Port	Maritime Data	μ_{sim}	σ_{sim}	$\mu_{sim} - 2\sigma_{sim}$	$\mu_{sim} + 2\sigma_{sim}$
Kokkola	1.7	3.6	1.4	0.7	6.4
Raahe	3.5	3.5	1.2	1.0	6.0
Oulu	3.8	3.0	1.6	−0.2	6.1
Kemi	2.3	3.1	1.0	1.1	5.1
Tornio	3.8	2.9	0.9	1.2	4.7

4.3.5. Sensitivity Analysis

Due to the high degree of complexity and randomness of the FSWNS, and the general lack of related data, the present simulation model rests on several assumptions. These concern the prevailing ice conditions, the IB transfer times, and the presence of pre-existing ice channels, among others. To assess the sensitivity of the simulation outcome to those assumptions, a sensitivity analysis is carried out. As per Figure 8, the sensitivity analysis indicates that the simulated number of instances of IB assistances is sensitive to variations in the assumed H_{eq}-values, but especially to assumptions concerning the presence of pre-existing ice channels. Specifically, by increasing or decreasing the assumed H_{eq}-values by 20%, the instances of assistance increase by 13% or decrease by 30%, respectively. By assuming that all those brash ice channels that as default are assumed to be present with a 50% probability (see Section 4.2.3) are present at all times, the instances of assistance decrease by 98%, whereas by assuming that the same ice channels are absent at all times, the instances of assistances increase by 45%.

The sensitivity analysis further indicates that the mean IB waiting time is sensitive to all the analyzed assumption variations. Specifically, by increasing or decreasing the assumed H_{eq}-values by 20%, the simulated mean waiting time for IB assistance increases by 6% or decreases by 23%, respectively. By increasing or decreasing the assumed IB transfer times by 50%, the simulated mean waiting time for IB assistance increases by 9% or decreases by 47%, respectively. By assuming that all those brash ice channels that as default are assumed to be present with a 50% probability (see Section 4.2.3) are present at all times, the mean waiting time decreases by 86%, whereas by assuming that the same ice channels are absent at all times, the waiting time increases by 69%.

4.3.6. Validation Summary

The above validation indicates that the proposed simulation approach works in principle. Regarding the number of ship arrivals per port, the simulation agrees well with the data. Regarding the number of instances of IB assistance and IB waiting times, the overall standard deviations between the averages of 35 independent simulation runs and the data are 13% and 18%, respectively. For each of the simulated metrics, the data fall within an approximated 95% confidence interval.

On average, the simulation appears to somewhat overestimate both the number of instances of assistance (+11%) and the IB waiting times (+7%). The overestimated number of assistances may indicate that, in the real world, ships are occasionally assisted over larger distances than the fixed IB operating areas assumed in the simulation. The overestimated mean IB waiting time, in turn, may in part be explained by the large individual peak values observed in Figure 7c. In the real world, such peak values can perhaps be avoided by factors not considered in the present simulation model, such as convoy operations including more than two ships and various active measures by the masters of the involved ships.

It should also be noted that the available maritime data, or the interpretation thereof, may include some errors. For instance, in individual cases, it is reported that a ship has received multiple instances of IB assistance during its inward voyage, but none during its outward voyage even though both voyages would have occurred in similar ice conditions. In addition, occasionally it appears like the data, instead of including all instances of IB assistance received by a ship on its way to or from a port, only include the last or the first instance of IB assistance before port arrival or after departure. Further studies are needed to identify and address such possible inconsistencies.

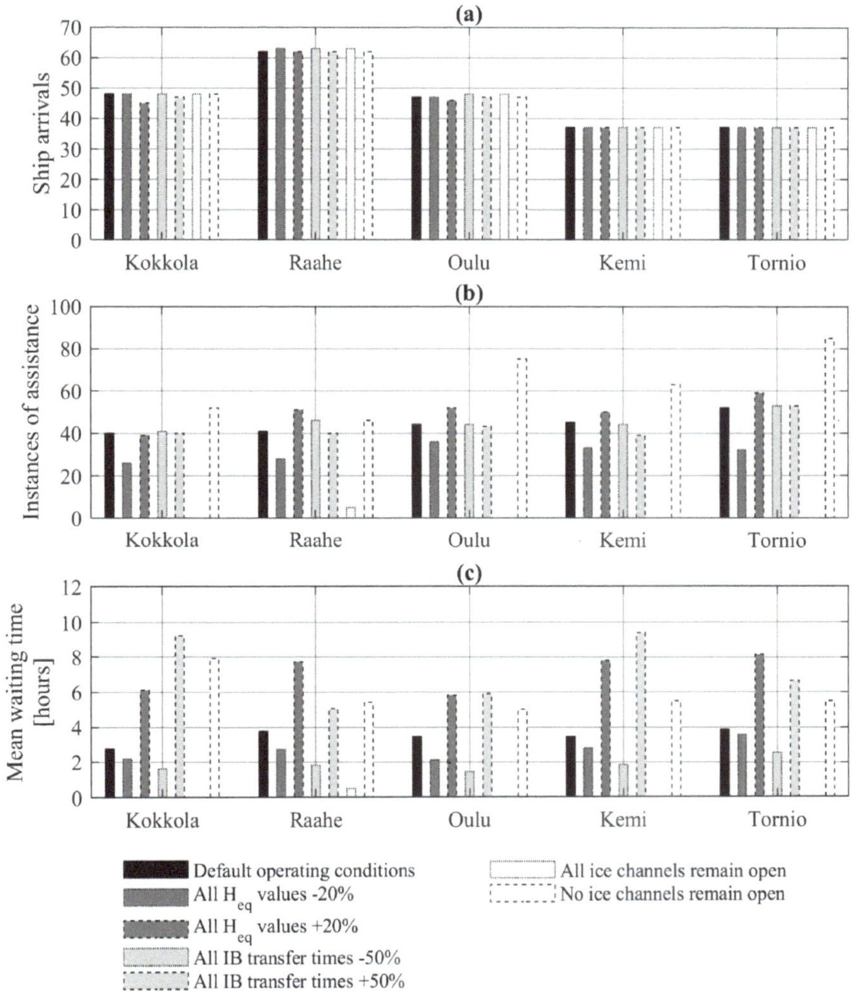

Figure 8. Sensitivity of the simulation output to changes in the underlying parameter values. (**a**) ship arrivals; (**b**) instances of assistance; (**c**) mean waiting times, the results were obtained by repeating the simulation so that one category of parameters was changed at a time, keeping all other parameter values constant.

5. Case Study: Impact of the EEDI on the Operating Performance of the FSWNS

As the EEDI is enforced on new ships only, the existing fleet of ships will only gradually be replaced by EEDI compliant ships. This case study is carried out for scenarios where either one-third (33%) or two-thirds (66%) of arriving ships, randomly selected, have been replaced by new EEDI compliant ships. It is further assumed that the achievable speed in ice of the new EEDI compliant ships is dependent on their size in DWT as per Figure 9. Accordingly, it is assumed that EEDI compliant ships are not able to operate independently in unbroken level ice. Additionally, as per the example presented in Figure 10, it is assumed that the speed of an EEDI compliant ship, when operating in a brash ice channel or with IB assistance, might be significantly lower than that of a corresponding non-EEDI ship. In the case studies, around 30% of the replaced vessels were in the category DWT < 5100, around 60% were in the category DWT 5100–15,000, and around 10% were in the category DWT 15,000–22,000.

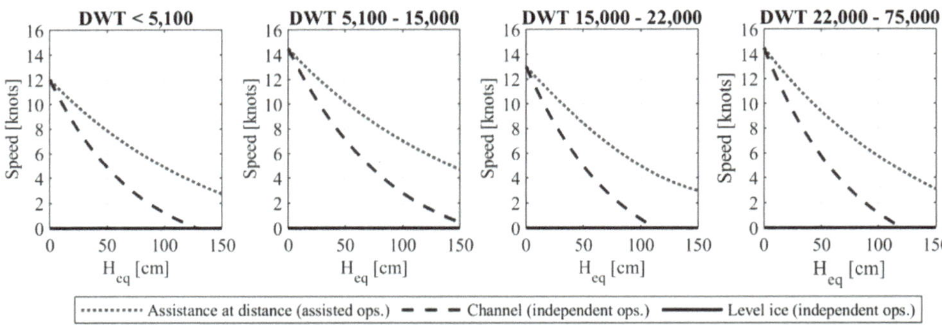

Figure 9. Assumed speed in ice of Energy Efficiency Design Index (EEDI) compliant ships [26].

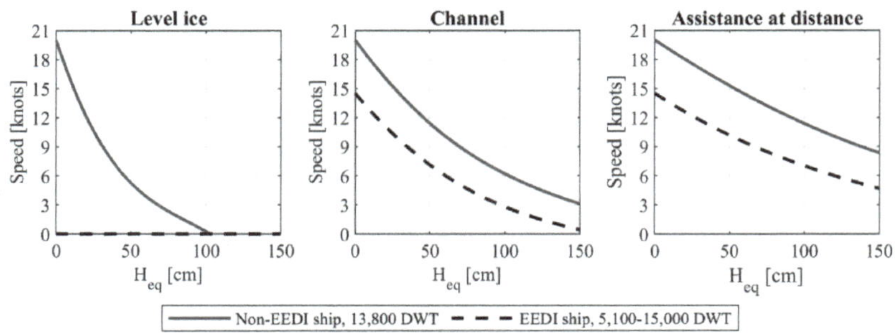

Figure 10. Example comparison of the assumed speed in ice of a non-EEDI ship and an EEDI-compliant ship [26].

The outcome of the simulated EEDI scenarios is presented in Figure 11. In the first scenario, in which one-third of the current fleet is replaced by EEDI-compliant ships, the number of cases of IB assistance is increased from 222 to 328 (+48%) and the total cumulated waiting times for IB assistance is increased from 775 h to 1378 h (+78%). In the second scenario, in which two-thirds of the current fleet is replaced with EEDI-compliant ships, the total number of cases of IB assistance is increased from 222 to 479 (+116%) and the total cumulated waiting times for IB assistance is increased from 775 h to 5157 h (+565%). In the second scenario, due to the extended waiting times for IB assistance, the total number of port arrivals decreases from 231 to 226 (−2%).

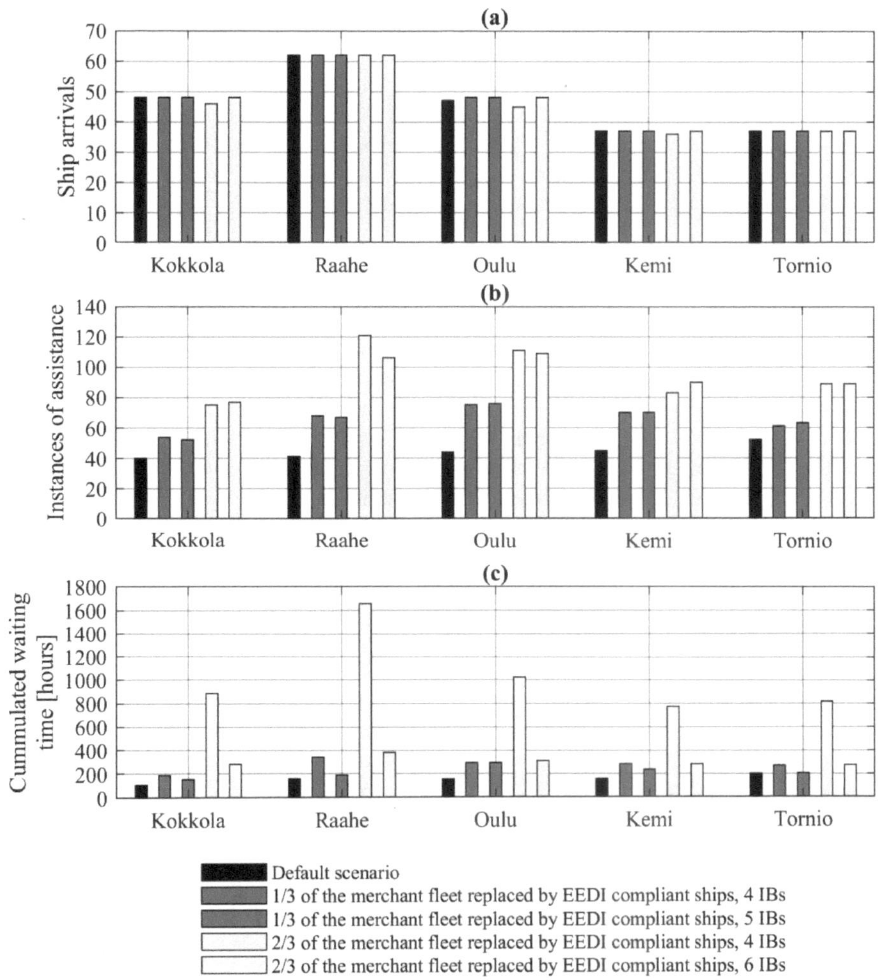

Figure 11. Influence of various EEDI scenarios on the operating performance of the Finnish–Swedish Winter Navigation System (FSWNS). (**a**) ship arrivals; (**b**) instances of assistance; (**c**) commulated waiting times, The results were obtained by repeating the simulation for the various fleet configurations considered, keeping all other parameter values constant. The cumulated port-specific waiting times reflect both the number of instances of assistance and the mean waiting time per instance

Additional simulations were carried out to assess whether the increase in IB waiting time from the assumed EEDI scenarios could be mitigated by increasing the number of IBs. As per Figure 11, the outcome from the simulations indicates that in the first and second EEDI scenarios, a significant increase in the cumulated IB waiting times can be mitigated by increasing the number of IBs from 4 to 5 and from 4 to 6, respectively. Notwithstanding, due to a lack of relevant data, this assessment is based on generalizing and conservative assumptions concerning the influence of the EEDI on the ice-going capability and other technical characteristics of the merchant fleet. Thus, this case study serves mainly as an example of how the presented approach can be applied to assess the performance of the FSWNS under various operating scenarios.

6. Discussion and Conclusions

This article presents a DES-based approach to predict the operating performance of the FSWNS under different operating scenarios. The approach is validated against real-world data on maritime traffic in the Bothnian Bay in the period 15 January–15 February 2010. In terms of the number of ship arrivals per port, representing the transport capacity of the FSWNS, the simulation agrees well with the data. In terms of the number of instances of IB assistance and IB waiting times, the standard deviations between the averages of 35 independent simulation runs and the data are 13% and 18%, respectively. For each of the simulated metrics, the data fall within an approximated 95% confidence interval. These findings indicate that the proposed DES-based approach can capture the complex behavior of the FSWNS and roughly estimate its operating performance.

Due to various identified knowledge and data gaps, as well as due to technical limitations of the applied simulation approach, the simulation model is based on generalized assumptions concerning convoy operations (a maximum of two ships are assisted by an IB at a time), IB transfer times (probabilistically determined), criteria for providing IB assistance (generalized criteria), and assumptions concerning the presence of brash ice channels (generalized assumptions), among others. A sensitivity analysis indicates that the simulated number of instances of IB assistance and IB waiting times are particularly sensitive to assumptions concerning the presence of brash ice channels. Future research is needed to address the limitations.

Considering the present limitations, the proposed approach appears best suited for scenario-based assessments in which the performance of the FSWNS is assessed for a limited period (e.g., one month) with heavy ice conditions during which the main shipping routes can be assumed fixed. The outcome of such a scenario-based assessment may indicate whether the capacity of the FSWNS under the simulated conditions is sufficient to keep the system in "balance".

Case studies were carried out in which the approach was applied to assess the impact of replacing various percentages of the present fleet of merchant ships entering the Bay of Bothnia by EEDI-compliant ships. In a scenario in which around one-third of the current fleet is replaced by EEDI-compliant ships, the simulation outcome indicates that the number of cases of IB assistance is increased from 222 to 328 (+48%) and that the cumulated waiting times for IB assistance is increased from 775 h to 1378 h (+78%). In another scenario in which around two-thirds of the current fleet is replaced with EEDI-compliant ships, the simulation outcome indicates that the total number of cases of IB assistance is increased from 222 to 497 (+116%) and that the cumulated waiting times for IB assistance is increased from 775 h to 5157 h (+565%). For the considered scenarios, simulation outcomes indicate that the predicted increase in IB waiting times can be largely mitigated if the number of IBs operating in the area is increased from 4 to 5 or from 4 to 6, respectively. However, due to a lack of detailed data on, e.g., how the EEDI would influence the ice-going capability and other technical characteristics of the affected merchant fleet, the outcome of the analysis is not conclusive.

In summary, the presented approach, which is one of the first attempts to simulate the FSWNS, may provide new insights into the behavior and performance of the FSWNS under different operating scenarios, considering a multitude of interlinked and self-reinforcing system behaviors. As such, the method may be used to support decision-making concerning the management of the FSWNS to meet future goals concerning operational efficiency. In the future, a potential further developed version of the model could be used for more holistic analyses, e.g., to analyze the cost- and energy-efficiency of the FSWNS, considering e.g., the impacts of climate change on sea ice conditions.

Author Contributions: M.B.: Conceptualization; Formal analysis; Funding acquisition; Investigation; Methodology; Software; Validation; Visualization; Writing—Original Draft. P.K.: Conceptualization; Data curation; Funding acquisition; Project administration; Writing—Review & Editing. All authors have read and agreed to the published version of the manuscript.

Funding: The authors gratefully acknowledge the funding by the Winter Navigation Research Board, a Finnish–Swedish cooperation co-funded by the Finnish Transport and Communications Agency, the Finnish Transport Infrastructure Agency, the Swedish Maritime Administration, and the Swedish Transport Agency.

Acknowledgments: We would like to thank the many organizations and individuals who provided guidance and support throughout this study. Special thanks to Morten Lindeberg for providing valuable input data.

Conflicts of Interest: The authors declare no conflict of interest.

Abbreviations

AMTS	Arctic maritime transport systems
DES	Discrete event simulation
DWT	Deadweight tonnage
EEDI	Energy Efficiency Design Index
FSICR	Finnish–Swedish ice class rules
FSWNS	Finnish–Swedish winter navigation system
GHGs	Greenhouse gasses
IB	Icebreaker
IMO	International maritime organization
MCR	Maximum continuous rating
PDF	Probability density function
PTT	Port turnaround time
SFC	Specific fuel consumption

Notations

C_F	Amount of CO_2 per gram of fuel (3.1144 g)
C_{max}	Maximum ice concentration
L	Navigation leg
H_{eq}	Equivalent ice thickness (cm)
H_{eq_avg}	Average equivalent ice thickness (cm)
H_{eq_max}	Maximum equivalent ice thickness (cm)
H_{max}	Maximum ice thickness (cm)
H_{min}	Minimum ice thickness (cm)
$H_{Ridging\_avg}$	Average equivalent ice thickness caused by ice ridging (cm)
$H_{Ridging\_max}$	Maximum equivalent ice thickness caused by ice ridging (cm)
v	Speed (knots)

Appendix A. Maritime Traffic Data

Table A1. List of ships arriving from Kvarken (Point A as per Figure 4).

Entry Time, Point B (h)	Ship Model	Port 1	Port 2	Port 3	PTT 1 (h)	PTT 2 (h)	PTT 3 (h)
1	I	Raahe			43		
1	A	Raahe			51		
1	P	Oulu	Kemi		5	4	
1	K	Kokkola			14		
5	N	Kokkola			48		
10	K	Kemi			100		
24	N	Tornio			46		
27	C	Oulu			14		
29	K	Kemi			60		
40	K	Raahe			61		
40	P	Kemi	Oulu		7	7	
48	C	Kemi	Kotka		21		
53	K	Tornio			77		
57	G	Kokkola	Raahe		26	61	
59	I	Kokkola			12		

Table A1. *Cont.*

Entry Time, Point B (h)	Ship Model	Port 1	Port 2	Port 3	PTT 1 (h)	PTT 2 (h)	PTT 3 (h)
63	K	Kokkola			31		
65	G	Raahe			89		
66	G	Kokkola			86		
69	M	Kokkola			20		
72	N	Kokkola			48		
74	K	Tornio			21		
76	F	Kokkola			73		
83	G	Raahe	Tornio		64	19	
91	C	Oulu			19		
91	G	Tornio	Kotka		25		
91	G	Kemi			28		
92	G	Kokkola	Kemi		60	53	
93	G	Kokkola			31		
95	O	Kemi	Oulu		4	12	
97	K	Tornio	Kemi		22	36	
99	K	Tornio			22		
101	C	Kokkola			50		
108	K	Raahe			69		
111	G	Raahe			51		
113	K	Kokkola			43		
119	K	Oulu			35		
122	K	Oulu			108		
125	G	Kokkola	Kemi	Oulu	35	73	65
126	K	Tornio			19		
133	G	Tornio			65		
134	A	Raahe			162		
135	K	Tornio	Raahe		22	30	
139	L	Kemi			51		
140	C	Oulu			19		
144	N	Oulu			34		
152	L	Oulu			17		
156	P	Oulu	Kemi		6	8	
157	G	Kemi			122		
157	G	Raahe			68		
158	K	Kokkola			27		
160	C	Kemi			30		
165	G	Raahe	Oulu		51	41	
167	N	Oulu			23		
172	K	Tornio			41		
178	K	Oulu			17		
182	F	Kokkola			32		
192	N	Raahe			52		
200	K	Kokkola			81		
200	K	Kokkola			76		
206	P	Oulu	Kemi		8	7	
207	J	Kokkola			64		
207	A	Raahe			57		
214	K	Kokkola	Raahe		15	48	
220	G	Tornio	Kokkola		50	55	
224	M	Raahe			34		
228	F	Kokkola			18		
228	K	Kokkola			97		

Table A1. *Cont.*

Entry Time, Point B (h)	Ship Model	Port 1	Port 2	Port 3	PTT 1 (h)	PTT 2 (h)	PTT 3 (h)
228	C	Oulu			-3		
228	C	Raahe	Pori		43		
230	G	Tornio			66		
235	C	Oulu			6		
235	C	Kokkola			7		
242	N	Oulu	Tornio		78	21	
243	I	Raahe			27		
244	G	Tornio	Raahe		20	49	
249	K	Tornio	Kemi		28	40	
250	K	Oulu			73		
260	B	Kokkola			54		
261	L	Kokkola			18		
262	O	Kemi	Oulu		7	7	
263	K	Tornio	Kotka		45		
266	P	Kemi			116		
268	M	Raahe			69		
275	G	Kokkola			64		
278	K	Tornio	Raahe		44	88	
293	G	Tornio			65		
298	C	Kokkola			24		
310	L	Oulu			18		
311	K	Oulu	Kemi		31	68	
319	K	Raahe			54		
321	K	Oulu	Pori		70		
323	P	Oulu	Kemi		10	11	
332	K	Kokkola			71		
336	A	Raahe			47		
337	B	Kokkola			24		
339	I	Kemi			17		
352	G	Kokkola			111		
360	N	Oulu			34		
364	K	Kokkola			21		
370	P	Kemi	Oulu		7	8	
374	M	Kokkola			22		
383	C	Kokkola			7		
384	N	Tornio			46		
393	N	Oulu	Tornio		104	23	
394	F	Raahe	Kokkola		41	65	
401	K	Tornio					
407	G	Kemi	Raahe		32	4	
419	E	Oulu			14		
419	A	Raahe			40		
424	G	Kokkola			94		
425	G	Tornio	Raahe		32	47	
426	P	Oulu			6		
434	B	Tornio	Raahe	Pori	147	29	
448	G	Tornio			79		
455	K	Kokkola			29		
457	A	Raahe			37		
470	J	Kokkola			64		
473	K	Kemi	Oulu		82	50	
474	K	Raahe			64		
481	K	Raahe			73		
486	B	Kokkola			17		

Table A1. *Cont.*

Entry Time, Point B (h)	Ship Model	Port 1	Port 2	Port 3	PTT 1 (h)	PTT 2 (h)	PTT 3 (h)
486	K	Tornio			29		
490	G	Kemi			39		
493	G	Oulu			84		
496	K	Oulu			15		
496	G	Tornio			77		
504	N	Raahe			52		
514	B	Kokkola			239		
514	P	Oulu	Kemi		9	7	
519	H	Kemi			145		
531	K	Raahe	Tornio		18	79	
536	P	Kemi	Oulu		8	10	
542	C	Oulu			22		
543	K	Raahe	Tornio		45	23	
554	G	Raahe	Oulu		25	58	
562	K	Raahe	Kokkola				
563	K	Tornio			47		
563	K	Raahe			8		
567	N	Oulu	Tornio		101	45	
574	K	Raahe			68		
576	G	Oulu			74		
576	N	Oulu			34		
577	G	Kemi	Tornio		44	23	
584	G	Tornio			67		
588	K	Kokkola			19		
593	L	Oulu			21		
593	K	Kemi	Oulu		52	45	
595	O	Kemi	Oulu		9	7	
596	H	Oulu					
610	I	Raahe			14		
621	A	Raahe			15		
626	C	Kokkola			49		
629	L	Kemi			17		
630	G	Tornio			65		
636	K	Tornio	Raahe		30	77	
648	N	Tornio			46		
659	G	Raahe			17		
660	M	Kokkola			28		
670	K	Tornio	Kokkola				
679	G	Kemi			51		
681	K	Kemi	Kokkola				
682	P	Oulu	Kemi		8	6	
683	L	Kokkola			15		
701	C	Kemi			15		
708	P	Kemi	Oulu		8	8	
710	F	Kokkola			64		
718	D	Raahe			92		
725	K	Raahe			7		
736	L	Oulu			29		
744	N	Oulu			34		
753	N	Oulu	Tornio				
754	K	Tornio			39		
755	F	Raahe	Kokkola		14	45	
764	G	Kokkola			32		
765	O	Kemi	Oulu				

Table A2. List of ships arriving from northern Sweden (Point B as per Figure 4).

Entry Time, Point B (h)	Ship Model	Port 1	Port 2	Port 3	PTT 1 (h)	PTT 2 (h)	PTT 3 (h)
31	A	Raahe			39		
67	A	Raahe			15		
110	A	Raahe			29		
137	A	Raahe			33		
155	M	Raahe			18		
164	A	Raahe			49		
249	A	Raahe			21		
371	A	Raahe			48		
388	M	Raahe			12		
455	A	Raahe			49		
455	M	Raahe			11		
541	A	Raahe			18		
593	A	Raahe			18		
608	A	Raahe			40		
645	A	Raahe			43		
686	A	Raahe			74		
737	M	Raahe			40		
758	A	Raahe			52		

Appendix B. Ice Data

Table A3. H_{eq_avg} (cm) determined by date and leg (L).

Date	L1	L2	L3	L4	L5	L6	L7	L8	L9	L10	L11	L12	L13	L14	L15
15 January 2010	6	14	32	13	29	24	32	30	23	32	30	20	33	33	25
16 January 2010	6	14	32	13	29	24	32	30	23	32	30	20	33	33	25
17 January 2010	6	14	32	13	29	24	32	30	23	32	30	20	33	33	25
18 January 2010	6	14	32	13	29	24	32	30	23	32	30	20	33	33	25
19 January 2010	6	14	32	13	29	24	32	30	23	32	30	20	33	33	25
20 January 2010	6	14	32	13	29	24	32	30	23	32	30	20	33	33	25
21 January 2010	6	14	32	13	29	24	32	30	23	32	30	20	33	33	25
22 January 2010	8	9	10	33	9	9	10	30	23	32	30	20	33	33	32
23 January 2010	8	9	10	33	9	9	10	30	23	32	30	20	33	33	32
24 January 2010	8	9	10	33	9	9	10	30	23	32	30	20	33	33	32
25 January 2010	8	9	10	33	9	9	10	30	23	32	30	20	33	33	32
26 January 2010	8	9	10	33	9	9	10	30	23	32	30	20	33	33	32
27 January 2010	8	9	10	33	9	9	10	30	23	32	30	20	33	33	32
28 January 2010	8	9	10	33	9	9	10	30	23	32	30	20	33	33	32
29 January 2010	4	5	5	33	5	5	5	38	21	32	38	21	38	38	43
30 January 2010	4	5	5	33	5	5	5	38	21	32	38	21	38	38	43
31 January 2010	4	5	5	33	5	5	5	38	21	32	38	21	38	38	43
1 February 2010	4	5	5	33	5	5	5	38	21	32	38	21	38	38	43
2 February 2010	4	5	5	33	5	5	5	38	21	32	38	21	38	38	43
3 February 2010	4	5	5	33	5	5	5	38	21	32	38	21	38	38	43
4 February 2010	4	5	5	33	5	5	5	38	21	32	38	21	38	38	43
5 February 2010	23	23	23	40	23	23	23	38	23	38	38	39	45	45	45
6 February 2010	23	23	23	40	23	23	23	38	23	38	38	39	45	45	45
7 February 2010	23	23	23	40	23	23	23	38	23	38	38	39	45	45	45
8 February 2010	23	23	23	40	23	23	23	38	23	38	38	39	45	45	45
9 February 2010	23	23	23	40	23	23	23	38	23	38	38	39	45	45	45
10 February 2010	23	23	23	40	23	23	23	38	23	38	38	39	45	45	45
11 February 2010	23	23	23	40	23	23	23	38	23	38	38	39	45	45	45
12 February 2010	41	33	33	40	33	33	33	45	33	45	45	39	55	55	52
13 February 2010	41	33	33	40	33	33	33	45	33	45	45	39	55	55	52
14 February 2010	41	33	33	40	33	33	33	45	33	45	45	39	55	55	52
15 February 2010	41	33	33	40	33	33	33	45	33	45	45	39	55	55	52

Table A4. H_{eq_max} (cm) determined by date and leg (L).

Date	L1	L2	L3	L4	L5	L6	L7	L8	L9	L10	L11	L12	L13	L14	L15
15 January 2010	20	20	45	20	30	36	36	40	30	30	40	30	40	40	39
16 January 2010	20	20	45	20	30	36	36	40	30	30	40	30	40	40	39
17 January 2010	20	20	45	20	30	36	36	40	30	30	40	30	40	40	39
18 January 2010	20	20	45	20	30	36	36	40	30	30	40	30	40	40	39
19 January 2010	20	20	45	20	30	36	36	40	30	30	40	30	40	40	39
20 January 2010	20	20	45	20	30	36	36	40	30	30	40	30	40	40	39
21 January 2010	20	20	45	20	30	36	36	40	30	30	40	30	40	40	39
22 January 2010	15	15	15	40	15	15	15	40	44	32	40	44	40	40	44
23 January 2010	15	15	15	40	15	15	15	40	44	32	40	44	40	40	44
24 January 2010	15	15	15	40	15	15	15	40	44	32	40	44	40	40	44
25 January 2010	15	15	15	40	15	15	15	40	44	32	40	44	40	40	44
26 January 2010	15	15	15	40	15	15	15	40	44	32	40	44	40	40	44
27 January 2010	15	15	15	40	15	15	15	40	44	32	40	44	40	40	44
28 January 2010	15	15	15	40	15	15	15	40	44	32	40	44	40	40	44
29 January 2010	5	5	5	40	5	5	5	50	30	30	50	30	50	50	54
30 January 2010	5	5	5	40	5	5	5	50	30	30	50	30	50	50	54
31 January 2010	5	5	5	40	5	5	5	50	30	30	50	30	50	50	54
1 February 2010	5	5	5	40	5	5	5	50	30	30	50	30	50	50	54
2 February 2010	5	5	5	40	5	5	5	50	30	30	50	30	50	50	54
3 February 2010	5	5	5	40	5	5	5	50	30	30	50	30	50	50	54
4 February 2010	5	5	5	40	5	5	5	50	30	30	50	30	50	50	54
5 February 2010	36	36	36	50	36	36	36	50	36	50	50	50	60	60	59
6 February 2010	36	36	36	50	36	36	36	50	36	50	50	50	60	60	59
7 February 2010	36	36	36	50	36	36	36	50	36	50	50	50	60	60	59
8 February 2010	36	36	36	50	36	36	36	50	36	50	50	50	60	60	59
9 February 2010	36	36	36	50	36	36	36	50	36	50	50	50	60	60	59
10 February 2010	36	36	36	50	36	36	36	50	36	50	50	50	60	60	59
11 February 2010	36	36	36	50	36	36	36	50	36	50	50	50	60	60	59
12 February 2010	54	46	46	50	46	46	46	60	46	59	60	59	70	70	64
13 February 2010	54	46	46	50	46	46	46	60	46	59	60	59	70	70	64
14 February 2010	54	46	46	50	46	46	46	60	46	59	60	59	70	70	64
15 February 2010	54	46	46	50	46	46	46	60	46	59	60	59	70	70	64

References

1. Finnish Shipowners Association. Key Figures of Maritime Transport in Finland. 2020. Available online: https://shipowners.fi/en/competitiveness/key-figures-of-maritime-in-finland/ (accessed on 15 April 2020).
2. Meriliitto. Finland is an Island. 2020. Available online: http://www.meriliitto.fi/?page_id=177 (accessed on 15 April 2020).
3. Finnish Transport and Communications Agency. *Statistics on International Shipping*; Finnish Transport and Communications Agency Traficom: Helsinki, Finland, 2018.
4. FMI. Ice Winter in the Baltic Sea. 2019. Available online: https://en.ilmatieteenlaitos.fi/ice-winter-in-the-baltic-sea (accessed on 15 September 2019).
5. Finnish Transport and Communications Agency. *Ice Class Regulations and the Application Thereof*; Finnish Transport and Communications Agency: Helsinki, Finland, 2017; Available online: https://www.finlex.fi/data/normit/43682/TRAFI_494_131_03_04_01_00_2016_EN_Jaaluokkamaarays_2017.pdf (accessed on 10 March 2020).
6. Jalonen, R.; Riska, K.; Hänninen, S. *Research Report No 57: A preliminary Risk Analysis of Winter Navigation in the Baltic Sea*; Winter Navigation Reseach Board/Helsinki University of Technology: Espoo, Finland, 2005.
7. ARCTIA. Homepage of Arctia Ltd. 2019. Available online: http://arctia.fi/en/ (accessed on 10 November 2019).
8. SJÖFARTSVERKET. Homepage of the Swedish Maritime Administration. 2019. Available online: https://www.sjofartsverket.se/en/Maritime-services/Winter-Navigation/Our-Icebreakers/ (accessed on 10 November 2019).
9. IMO. Energy Efficiency Measures. International Maritime Organization. Available online: http://www.imo.org/en/OurWork/Environment/PollutionPrevention/AirPollution/Pages/Technical-and-Operational-Measures.aspx (accessed on 10 June 2019).
10. Bergström, M.; Erikstad, S.O.; Ehlers, S. A simulation-based probabilistic design method for arctic sea transport systems. *J. Mar. Sci. Appl.* **2016**, *15*, 349–369.
11. Craig, D. Discrete-Event Simulation. Memorial University of Newfoundland: St. John's, NL, Canada, 1996.

12. Lindeberg, M.; Kujala, P.; Sormunen, O.-V.E.; Karjalainen, M.; Toivola, J. Simulation model of the Finnish winter navigation system. In Proceedings of the 13th International Marine Design Conference (IMDC 2018), Helsinki, Finland, 10–14 June 2018.
13. Finnish Transport and Communications Agency. *Ice Classes of Ships*; Finnish Transport and Communications Agency: Helsinki, Finland, 2019.
14. Baltic Icebreaking Management. *Baltic Sea Icebreaking Report 2014–2015*; Baltic Icebreaking Management: Helsinki, Finland, 2015.
15. International Maritime Organization. *2018 Guidelines on the Method of Calculation of the Attained Energy Efficiency Design Index (EEDI) for New Ships*; International Maritime Organization: London, UK, 2018.
16. International Maritime Organization. *Train the Trainer (TTT) Course on Energy Efficient Ship Operation. Module 2—Ship Energy Efficiency Regulations and Related Guidelines*; International Maritime Organization: London, UK, 2016.
17. Kämäräinen, J. *EEDI -Hanke ja Laivamittauskampanja. Merenkulun Tutkimusseminaari*; Finnish Transport Safety Agency: Helsinki, Finland, 2017.
18. Prime Minister´s Office. *The impact of International Maritime Climate and Environmental Regulation on the Finnish Economy*; Prime Minister´s Office: Helsinki, Finland, 2017.
19. MathWorks. SimEvents. Available online: https://se.mathworks.com/help/simevents/index.html?s_cid=doc_ftr (accessed on 1 September 2020).
20. Goerlandt, F.; Montewka, J.; Zhang, W.; Kujala, P. An analysis of ship escort and convoy operations in ice conditions. *Saf. Sci.* **2017**, *95*, 198–209. [CrossRef]
21. Riska, K. *Ship–Ice Interaction in Ship Design: Theory and Practice*; Encyclopedia of Life Support Systems (EOLSS): Paris, France, 2010.
22. Kuuliala, L.; Kujala, P.; Suominen, M.; Montewka, J. Estimating operability of ships in ridged ice fields. *Cold Reg. Sci. Technol.* **2016**, *135*. [CrossRef]
23. Bergström, M.; Erikstad, S.O.; Ehlers, S. The Influence of model fidelity and uncertainties in the conceptual design of Arctic maritime transport systems. *Ship Technol. Res. Schiffstechnik* **2017**, *64*, 40–64. [CrossRef]
24. Sormunen, O.-V.; Berglund, R.; Lensu, M.; Kuuliala, L.; Li, F.; Bergström, M.; Kujala, P.; Linna, M. Comparison of vessel theoretical ice speeds against AIS data in the Baltic Sea. In Proceedings of the 13th International Marine Design Conference (IMDC 2018), Helsinki, Finland, 10–14 June 2018.
25. Li, Z.; Ringsberg, J.W.; Rita, F. A voyage planning tool for ships sailing between Europe and Asia via the Arctic. *Ships Offshore Struct.* **2020**. [CrossRef]
26. WINMOS II. *Winter Navigation Motorways of the Sea II (WINMOS II)*; WINMOS II Consortium: Helsinki, Finland, 2017; Available online: http://www.winmos.eu/ (accessed on 10 March 2020).
27. SMHI. Sea Ice. Archived Charts and Reports. Available online: http://www.smhi.se/oceanografi/istjanst/havsis_en.php (accessed on 15 October 2019).
28. Kankaanpää, P. Distribution, morphology and structure of sea ice pressure ridges in the Baltic Sea. *Int. J. Geogr.* **1997**, *175*, 139–240.
29. United States Coast Guard. *The Engineer's Digest*; United States Coast Guard: Washington, DC, USA, 1946.
30. Montewka, J.; Goerlandt, F.; Kujala, P.; Lensu, M. Towards probabilistic models for the prediction of a ship performance in dynamic ice. *Cold Reg. Sci. Technol.* **2015**, *112*, 14–28. [CrossRef]

© 2020 by the authors. Licensee MDPI, Basel, Switzerland. This article is an open access article distributed under the terms and conditions of the Creative Commons Attribution (CC BY) license (http://creativecommons.org/licenses/by/4.0/).

applied sciences

MDPI

Article

A Framework for Integrating Life-Safety and Environmental Consequences into Conventional Arctic Shipping Risk Models

Thomas Browne [1,2,*], Rocky Taylor [1], Brian Veitch [1,*], Pentti Kujala [3], Faisal Khan [1] and Doug Smith [1]

1 Faculty of Engineering and Applied Science, Memorial University of Newfoundland, St. John's, NL A1B 3X5, Canada; rstaylor@mun.ca (R.T.); fikhan@mun.ca (F.K.); d.smith@mun.ca (D.S.)
2 Ocean, Coastal and River Engineering Research Centre, National Research Council of Canada, St. John's, NL A1B 3T5, Canada
3 Department of Mechanical Engineering, Aalto University, 02150 Espoo, Finland; pentti.kujala@aalto.fi
* Correspondence: thomas.browne@mun.ca (T.B.); bveitch@mun.ca (B.V.)

Received: 1 April 2020; Accepted: 21 April 2020; Published: 23 April 2020

check for updates

Featured Application: A risk assessment framework that supports Arctic voyage planning and real-time operational decision-making through assignment of operational criteria based on the likelihood of ice-induced damage and the potential consequences.

Abstract: The International Code for Ships Operating in Polar Waters (Polar Code) was adopted by the International Maritime Organization (IMO) and entered into force on 1 January 2017. It provides a comprehensive treatment of topics relevant to ships operating in Polar regions. From a design perspective, in scenarios where ice exposure and the consequences of ice-induced damage are the same, it is rational to require the same ice class and structural performance for such vessels. Design requirements for different ice class vessels are provided in the Polar Code. The Polar Operational Limit Assessment Risk Indexing System (POLARIS) methodology provided in the Polar Code offers valuable guidance regarding operational limits for ice class vessels in different ice conditions. POLARIS has been shown to well reflect structural risk, and serves as a valuable decision support tool for operations and route planning. At the same time, the current POLARIS methodology does not directly account for the potential consequences resulting from a vessel incurring ice-induced damage. While two vessels of the same ice class operating in the same ice conditions would have similar structural risk profiles, the overall risk profile of each vessel will depend on the magnitude of consequences, should an incident or accident occur. In this paper, a new framework is presented that augments the current POLARIS methodology to model consequences. It has been developed on the premise that vessels of a given class with higher potential life-safety, environmental, or socio-economic consequences should be operated more conservatively. The framework supports voyage planning and real-time operational decision making through assignment of operational criteria based on the likelihood of ice-induced damage and the potential consequences. The objective of this framework is to enhance the safety of passengers and crews and the protection of the Arctic environment and its stakeholders. The challenges associated with establishing risk perspectives and evaluating consequences for Arctic ship operations are discussed. This methodology proposes a pragmatic pathway to link ongoing scientific research with risk-based methods to help inform recommended practices and decision support tools. Example scenarios are considered to illustrate the flexibility of the methodology in accounting for varied risk profiles for different vessel types, as well as incorporating input from local communities and risk and environmental impact assessments.

Keywords: arctic shipping; POLARIS; risk assessment; consequence modeling; life-safety; environmental safety; operations management

1. Introduction

Ships and crews operating in Arctic and Antarctic environments are exposed to a number of unique risks. The presence of sea ice and icebergs can impose additional loads on the hull, propulsion system, and appendages of a vessel. Cold temperatures, poor weather, and marine icing may reduce the effectiveness of components of the ship, ranging from deck machinery and emergency equipment to sea suctions. A relative lack of good charts, communication systems, and other navigational aids in the Polar regions, and the remoteness of these areas, makes rescue and clean-up operations difficult and costly.

Arctic maritime operations are complex socio-technical systems [1–4]. Not only are the vessels and crews exposed to risks, but the environment and local communities will be impacted by the consequences of shipping. In this regard, there are socio-economic aspects to Arctic shipping that should be appreciated. Holistic risk assessment frameworks and operational decision support tools should account for the needs and interests of the diverse stakeholders within these regions.

The Polar Code [5], adopted by the International Maritime Organization (IMO) and entered into force in 2017, addresses many of the design and operational challenges associated with marine transportation in the Arctic and Antarctic. It provides guidance on ship design, construction, equipment, operations, training, search and rescue, and environmental protection. The Polar Code also provides the Polar Operational Limit Assessment Risk Indexing System (POLARIS), an operational decision support tool that provides guidance on the operational limits of a vessel as a function of the vessel's ice class and observed or forecasted ice conditions [6].

The Polar Code was created, in part, in response to recommendations from the Artic Council's 2009 Arctic Marine Shipping Assessment report [7,8]. Recommendations include the need to enhance Arctic marine safety and the need to protect Arctic people and the environment.

A recent study of the negative impacts of ship activity on Arctic marine mammals notes a lack of maritime guidance for the management of environmental impacts of Arctic shipping [9]. For example, operational decision-support tools should account for region-specific environmental vulnerabilities to oil spills [10].

The current POLARIS methodology accounts for the likelihood of a vessel to incur ice-induced damage, but it does not properly account for the potential consequences resulting from the ice-induced damage event. The operational limitations for a vessel in ice should be assessed based on a risk profile that incorporates life and environmental consequences.

The risk-based design methods employed in the International Standard for Arctic Offshore Structures, 19906 [11] provide an approach to the treatment of life and environmental safety. A similar approach may be adapted for ships operating in ice environments. Such an approach would explicitly consider the number of persons on board (POB), the cargo being transported, regional aspects, and the operational exposure with respect to life-safety, environmental, and socio-economic consequences.

It is noted that while significant progress has been made in developing probabilistic ice load models to link vessels' ice exposure, extreme ice loads, and ice class selection [12–16], design aspects are outside the scope of the operationally focused work presented here.

The methodology presented in this paper draws from both the Polar Code and the International Organization for Standardization (ISO) 19906 to explore ways in which life-safety and environmental safety considerations employed in ISO 19906 for offshore structures can be applied to ice class ships operating in the Arctic and Antarctic regions.

A new risk assessment framework is presented that augments the current POLARIS methodology to model consequences. The framework accounts for the likelihood of incurring ice-induced damage and the potential life-safety, environmental, and socio-economic consequences.

The benefit of the proposed framework is that it builds on the current POLARIS methodology, providing operational guidance considering the potential severity of consequences. When the perceived risk of operating in ice increases, additional operational restrictions are imposed to maintain an equivalent safety level. The objective of the framework is to enhance Arctic marine safety and the protection of the Arctic environment and its stakeholders by supporting operational decision-making for ships operating in Polar regions.

The current POLARIS methodology was published as "interim guidance", to be updated based on experience gained after several years of use [6]. The proposed framework can be seen as a recommended modification to the current POLARIS methodology.

Section 2 compares the design code philosophies of the Polar Code and ISO 19906. Section 3 reviews the challenges associated with evaluating life-safety and environmental consequences of Arctic shipping. Section 4 introduces the proposed life-safety and environmental consequence framework. Section 5 provides scenarios to illustrate the flexibility of the methodology in accounting for varied risk profiles and different vessel types and cargos. In Section 6, the framework is applied to a benchmark case study incorporating data from published Arctic marine shipping assessments. Section 7 discusses the merits of the proposed framework and areas for future work. Section 8 is the conclusion.

2. Background

2.1. Design Code Philosophy: Ships vs. Structures

The goal of ice class ship design rules is to provide a vessel design that satisfies specified standards. The vessel design, including structure, propulsion systems, and auxiliary systems, is assessed against a range of specific conditions, such as ice, low temperatures, and high latitude, as well as the potential need to abandon the ship onto ice or land. As a result, ice class rules tend to be more prescriptive than performance-based. Vessel class is selected to satisfy the operational profile specified by the owner and the owner/operator is then responsible for safely operating the vessel within the bounds of its capabilities.

In comparison, offshore structure codes use a risk-based approach focused on ensuring target safety levels are achieved. The reliability of the asset depends, in part, on exposure levels, which are determined based on an assessment of the potential life-safety, environmental, and economic consequences associated with a particular installation. The structure is designed to safely withstand the site-specific environmental conditions and other operational requirements that it is expected to encounter over its design life. Support activities, such as ice management, may be carried out as part of routine operations to help ensure safety.

Presently, there is no direct account for life safety class and environmental safety class in shipping codes. Adopting exposure levels, similar to those used in the design of offshore structures, is a rational approach for incorporating life-safety, environmental, and socio-economic consequences in the operational decision making of ice class vessels.

2.2. IMO Regulations

The International Convention for the Safety of Life at Sea (SOLAS) [17] promotes safety of life at sea through design and construction, requirements for onboard lifesaving appliances, and operational guidelines and restrictions. Structurally, a ship is considered safe if it has sufficient strength, integrity, and stability. Operationally, communication, planning, and procedures play important roles in life safety and safe navigation, including regulations and guidelines addressing voyage planning, ships' routing and reporting system requirements, and vessel traffic services.

The International Convention for the Prevention of Pollution from Ships (MARPOL) [18] promotes the prevention of operational and accidental pollution from ships. Pollution by oil and other substances as a result of a marine accident is primarily mitigated through structural design and equipment requirements. Certain ship types may have more stringent design requirements depending on the type and quantity of cargo. Operational requirements primarily focus on controlling pollution from intentional, operational discharges or routine operations such as ship-to-ship transfer of crude oil.

From an operational risk management perspective, the procedures prescribed in SOLAS and MARPOL are broadly applied across a range of ship types.

2.3. Polar Code and POLARIS

The Polar Code was developed, in part, to address the demands associated with the operation and navigation of ships in Polar regions that are not sufficiently captured in the existing requirements outlined in SOLAS and MARPOL. The Polar Code covers the full range of design, construction, equipment, operations, training, search and rescue, and environmental protection matters relevant to ships operating in the inhospitable waters surrounding the two poles.

Vessels being designed for operation under the Polar Code are required to undergo an operational assessment to establish the vessels' operational capabilities and limitations. The operational assessment follows the risk-based IMO Formal Safety Assessment [19], which forms the basis for a vessel's Polar Ship Certificate and Polar Waters Operational Manual (PWOM), and implicitly incorporates crew and environmental safety. In this regard, the Polar Code works well for vessel design and class selection.

For voyage planning and real-time decision making on the bridge, the POLARIS methodology assesses the operational limitations of an ice class vessel. POLARIS was developed, in part, from experiences gained through use of Canada's Arctic Ice Regime Shipping Systems (AIRSS) and the Russian Ice Certificate. POLARIS evaluates the risks posed to a ship operating in ice based on assigned ice class and the ice regime. The ice regime may be historic or forecasted in the case of voyage planning, or it may be observed from the bridge of the ship in the case of real-time decision-making.

For a given vessel class, POLARIS assigns Risk Index Values (RIVs) corresponding to each ice type, where a given ice regime can be comprised of several different ice types. The total Risk Index Outcome (RIO) is determined by the summation of the RIVs for each ice type present in the ice regime multiplied by the corresponding concentration of that ice type (expressed in tenths), as shown in Equation (1):

$$\mathrm{RIO} = (\mathrm{C1} \times \mathrm{RIV1}) + (\mathrm{C2} \times \mathrm{RIV2}) + \cdots + (\mathrm{Cn} \times \mathrm{RIVn}) \tag{1}$$

where C1 ... Cn = concentration (in tenths) of each ice type within the ice regime and RIV1 ... RIVn = the corresponding RIVs for each ice type.

The calculated RIO governs the operational criteria for the vessel: 'Normal operation' ($\mathrm{RIO} \geq 0$), 'elevated operational risk' ($-10 \leq \mathrm{RIO} \leq 0$), or 'operation subject to special consideration' ($\mathrm{RIO} < -10$). Response measures for 'elevated operational risk' include reducing speed, additional watch keeping, or icebreaker escort, while 'operation subject to special consideration' measures include further reduction of speed, course alteration, or other special measures to reduce risk. Guidance on procedures for operational criteria should be documented in the vessel's PWOM.

The POLARIS methodology was recently validated as a suitable means for assessing the risk of structural damage of ice-going vessels [12]. Two vessels were instrumented to record full-scale ice-induced hull loads and ice concentrations during their voyages. POLARIS was used to determine the optimal ice class to allow navigation in both scenarios. Optimal ice class was also evaluated based on the required hull strength to mitigate the risk of structural damage. For each scenario, POLARIS identified the same optimal ice class as the structural risk analysis.

Despite the Polar Code promoting a holistic approach to risk management, POLARIS is not a single solution for operational risk management; it only accounts for the risk of structural damage. There is no consideration for the potential consequences of a ship damaged by ice. Operators require complementary tools and additional data to support a more holistic, risk-based decision-making process [20].

The intent of POLARIS is that the operational criteria for a vessel in a given ice regime corresponds to the operational capabilities of the vessel's ice class. In effect, a high ice class vessel operating in heavy ice will have a similar perceived risk level as a non-ice strengthened vessel in open water [21]. However, there is significant uncertainty in estimated ice loads for different ice–ship interaction scenarios [12]. POLARIS does not guarantee safe navigation and incidents can still occur. Life-safety and environmental consequences that can result from ice damage to a vessel need to be accounted for in operational decision-making.

2.4. ISO 19906 Life-Safety Classes, Consequences, and Exposure

The risk-based approach employed by ISO 19906 for the design of Arctic offshore structures specifies that the reliability of a structure should reflect its exposure level with respect to life-safety, environmental, and economic consequence categories. For a given exposure level, extreme and abnormal level environmental loads corresponding to specified exceedance probabilities are determined. Calibrated action and material/resistance factors are then applied to determine the design actions corresponding to structural limit states specified in the code.

In ISO 19906, the life-safety category of an asset takes into consideration the safety of personnel and the probability of a safe evacuation. Three life-safety categories are defined for Arctic offshore structures: S1 (manned non-evacuated), S2 (manned evacuated), S3 (unmanned).

Similarly, the consequence category of an asset takes into consideration the potential risks in relation to the safety of personnel responding to an incident, environmental damage, and economic loss. Three consequence categories are defined in ISO 19906 for Arctic offshore structures: C1 (high consequence), C2 (medium consequence), C3 (low consequence).

The exposure level of an asset is then determined as a function of the assessed life-safety and consequence categories. Table 1 is used to determine the exposure level as a function of life-safety categories and consequence categories [11].

Table 1. Determination of exposure level (based on International Organization of Standardization (ISO) 19906 [11]).

Life-Safety Category	Consequence Category		
	C1 High	C2 Medium	C3 Low
S1: manned non-evacuated	L1	L1	L1
S2: manned evacuated	L1	L2	L2
S3: unmanned	L1	L2	L3

Manned non-evacuated (S1) refers to Arctic offshore structures in which there is no planned evacuation of personnel prior to a forecasted design environmental loading event. Manned evacuated (S2) refers to a platform in which evacuation of personnel prior to a forecasted design environmental loading event is planned. Unmanned (S3) refers to a platform that is not normally manned [11].

For offshore structures, the life-safety and consequence categories and corresponding exposure levels are defined during the design process and influence the structural capacity of the design to achieve target safety levels. The application proposed here is for management of operational risk of Arctic ships. When the perceived risk of operating a vessel in ice is increased, additional operational restrictions are required to maintain an equivalent risk level.

2.5. Linking Risk-Based Approach and Polar Rules

Concerning Arctic marine transport, the number of people on board, the amount and type of potential pollutants being transported, and the characteristics of the operating region directly impact a vessel's risk exposure and the severity of consequences. It is logical to account for these higher and lower risk levels in the rules. From an operational standpoint, an approach that explicitly reflects life-safety, environmental, and socio-economic consequences in a risk-based framework is needed.

The Polar Code works well for vessel design and class selection, but a more explicit approach that reflects life-safety and environmental/socioeconomic consequences in a risk-based framework is needed. A risk-based framework can be linked to the operational limitations of the vessel to support decision-making (i.e., POLARIS).

3. Arctic Shipping Risk

3.1. Life-Safety and Environmental Consequences

Evaluating life-safety and environmental consequences related to Arctic shipping risks is challenging. A lack of accident data and experience limits the application of conventional risk approaches that rely on empirical event probabilities and quantified consequence severities. Alternative, unconventional risk assessments are necessary.

Marchenko et al. [22] used qualitative, expert-based risk analysis of ship accidents to establish risk levels for a range of vessel types and incidents in various regions of the Arctic. They established that the perceived likelihood of high consequence events increases with increasing vessel traffic, the number of passengers, and the presence of hazardous cargos. They highlighted that the severity of life-safety and environmental consequences can escalate in the Arctic due to a lack of emergency response resources and the harsh environmental conditions.

Oil spills are the dominant threat posed by Arctic shipping [8] but evaluating environmental consequences is complex. The presence of dangerous goods onboard a vessel introduces the risk of environmental damage, but the consequence severity is not simply a product of accident potential and oil spill trajectory [23]. An evaluation of environmental risk from an oil spill should consider habitat exposure areas, recovery potentials of species and habitats, and the current state of the habitats [23,24]. Nevalainen et al. [25] go further, suggesting that risk assessments must consider the entire ecosystem (rather than species-specific) to identify long-term impacts and provide a holistic understanding of the impacts of an oil spill. Given the complexity of Arctic oil spills, evaluating oil spill risk requires multidisciplinary expert knowledge and region-specific analyses.

Evaluating life-safety consequences of Arctic shipping is also complex, with many dynamic factors (spatial and temporal). The number of passengers and crew on board a vessel and the ability to mount safe and effective escape, evacuation, and rescue (EER) will influence the severity of potential life-safety consequences. The rescue of crew and passengers is daunting: Limited regional search and rescue capabilities, scarce and aging infrastructure, long response times, and inadequate emergency response capacities for large-scale incidents (e.g., large cruise vessels present a high life-safety risk in the Arctic) [22,26,27].

Risk-based operational decision-making for ice class vessels should be based on a careful consideration of all consequences and the integration of multidisciplinary knowledge. Evaluating the severity of environmental and life-safety consequences is complex with many dynamic factors. The framework proposed here integrates multidisciplinary knowledge for scenario-based risk management for ships operating in ice.

3.2. Risk Perspectives and Applications

Different risk perspectives and applications have implications on risk acceptance and operational decision making. Aven et al. [28] remind us that there is a broad range of complex risk perspectives, and risk-based decision-making should aim to incorporate the full range of stakeholders and their diverse perspectives on risk and consequence (e.g., scientific, economic, social, and cultural).

Goerlandt and Montewka [29] examined a range of risk definitions and perspectives that have been applied in maritime transportation. Based on this, it can been seen that the risk definition adopted in POLARIS accounts for the likelihood of an undesirable event (i.e., the vessel incurring ice damage), but does not account for the relevant consequences of that event. The framework proposed here aims to complement POLARIS by accounting for the severity of consequences resulting from an ice damage event.

Similar foundational issues are present in oil spill risk analysis. Parviainen et al. [30] provided context on the ambiguity in risk perspectives and risk governance related to oil spills in the Barents Sea. Through the development of qualitative mental models for various stakeholders, they demonstrated there are multiple ways in which stakeholders define and understand risk, but existing risk assessment and management practices do not reflect this broad range of perspectives.

Further adding to the complexity of assessing Arctic shipping risk is the treatment of uncertainty, or strength of evidence. Risk analysis for maritime transportation seldom incorporates an assessment of uncertainty [1,29] despite the implications it has on risk acceptance and decision making.

The lack of experience and data for Arctic operations has made expert elicitation a common approach to risk. Expert judgement introduces additional uncertainties and bias that need to be considered and communicated to decision makers [30,31].

There are also "black swan" events [4,22,32]. These are rare or surprising events with the potential for severe or extreme consequences that are not captured in traditional risk analyses. As Arctic shipping activity increases and high risk exposure vessels enter new geographic regions, "black swan" events should be considered.

Several recent studies have proposed operational risk frameworks for ships in ice with a variety of risk perspectives and applications. Bayesian networks are a common risk assessment methodology applied to Arctic shipping. They allow the integration of quantitative and qualitative (e.g., expert knowledge) data and a means of quantifying uncertainty. Montewka et al. [33] used empirical data sets and Bayesian networks to assess ship performance (speed) as a function of ice conditions. Such a model supports operational risk management to avoid besetting in ice and to manage fuel economy. Fu et al. [34] adopted a Bayesian Belief Network using empirical data supplemented with expert judgement to assess the risk influencing factors leading to ship besetting.

Bergstrom et al. [35] investigated goal- and risk-based design to assess the performance of ships operating in ice. The ship is treated as a subcomponent of a larger Arctic marine transport system, utilizing principles of system-based design. While their intent was to incorporate a risk-based assessment of system performance at the design stage, their systems thinking approach has merit for the scenario-based framework proposed here.

Smith et al. [2] used the Functional Resonance Analysis Method (FRAM) to model Arctic ship navigation as a complex system and analyze the system functions (human, technical, and organizational) that influence ship performance. While FRAM is not a risk assessment methodology, it promotes a holistic understanding of system dynamics that can support real-time risk-based decision making.

Figure 1 summarizes the range of factors and foundational issues that should to be considered in the evaluation of consequences of Arctic shipping.

The risk framework proposed here aims to establish a risk perspective that captures the needs and interests of the diverse stakeholders of Arctic shipping, and to move towards more holistic risk management practices.

Figure 1. The factors and foundational issues that influence the evaluation of Arctic shipping consequences.

4. Proposed Life-Safety and Environmental Consequence Framework

The premise of the proposed life-safety and environmental consequence framework is that if you have two ships with the same ice class in the same ice conditions, a vessel having higher potential consequences for life-safety and/or environmental safety should be operated more conservatively. While two scenarios may have the same structural ice risk, when consequence severities are considered, the overall risk may be higher or lower and this should be reflected in the required operating limits.

Figure 2 shows the chain of consequences considered in the risk framework presented here. In the event the structural capacity of a ship's hull is exceeded, operational intervention measures will be employed to mitigate consequences. Should these measures be inadequate, EER may be required, which may have life-safety impacts, and there is the potential for a spill, which may have ecological and socio-economic impacts.

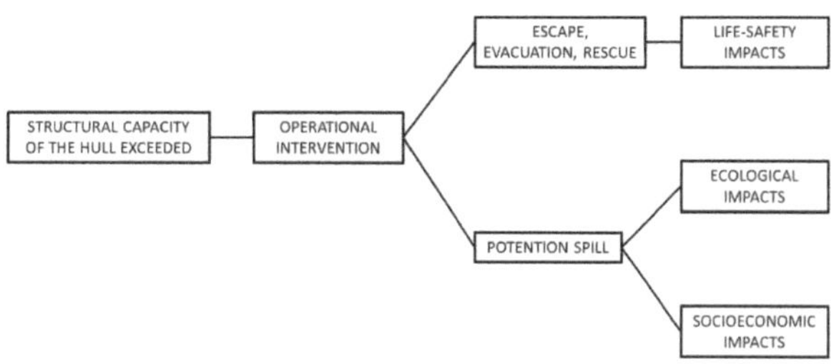

Figure 2. The chain of consequences following ice damage to a vessel considering life and environmental safety.

4.1. Overview of Methodology

Ship design, class selection, and performance criteria already follow well-established methodologies that are defined in the Polar Class rules. The approach proposed here is to add an exposure adjustment term in the POLARIS methodology that reflects the higher consequence operations. It will also provide a mechanism to recognize measures taken by vessel owners and operators for reducing risk. This is relevant for unmanned/autonomous vessels that do not carry pollutants, as such vessels could be operated more aggressively in a given ice regime with no impact to life-safety or the environment.

Following a similar approach to that used in ISO 19906, a life-safety category and an environmental/socio-economic consequence category is used to inform the assessment of a vessel's exposure level. The exposure level corresponds to an RIV adjustment factor, similar to the current approach in POLARIS for the treatment of seasonal ice decay. In doing so, the proposed framework guides the formulation of a RIO corresponding to the magnitude of life-safety, environmental, and socio-economic risks and consequences.

4.2. Life-Safety Categories

The proposed risk assessment starts with identification of the life-safety category of a ship, which is a ranking that reflects its exposure in relation to the safety of crew and passengers. It could also reflect the response plan adopted by the vessel and emergency response capacities along the planned route. As an example, the life-safety categories may be divided into four ranges based on POB, as defined in Table 2.

Table 2. Life-safety categories.

Life-Safety Category	Persons on Board (POB) Range
S1: high life-safety	POB > 500
S2: moderate life-safety	50 < POB ≤ 500
S3: low life-safety	0 < POB ≤ 50
S4: unmanned / autonomous	POB = 0

These life-safety categories are not equivalent to those provided in ISO 19906. ISO 19906 assesses life-safety for Arctic offshore structures based on site-specific, risk-based, designed EER strategies. It is recognized that factors influencing life-safety for ships transiting the Arctic (e.g., emergency response capacities and times, and environmental conditions) will vary spatially and temporally. The life-safety categories in Table 2 are intended to reflect the scale of search and rescue response operations required to assist in an emergency. More emergency response resources are required to ensure a safe response for vessels with higher numbers of POB. The categories provided here are used for illustrative purposes.

4.3. Environmental/Socio-Economic Consequence Categories

The next aspect of the proposed method is assessment of the environmental and socio-economic consequence categories associated with the vessel and its planned route. Consequences to be considered may be grouped into region-specific sensitivities, as well as vessel-specific considerations relating to the amount and type of potential pollutants.

Regulators and government agencies will be responsible for developing policies for Arctic maritime safety. Policy decisions will need to be informed by many different types of knowledge, such as multidisciplinary risk analyses, stakeholder engagements, and collaborative mapping. While a detailed discussion of risk-based policy development is beyond the scope of the present work, there are important links to the methodology proposed here. It is possible for environmental risk information to be communicated through geo-spatial maps, similar to those in the Arctic Council's report on the identification of Arctic marine areas of heightened ecological and cultural significance [24]. Such maps can be used to inform operational decision making and route planning.

Proposed approaches for capturing and categorizing these different types of consequences are described below.

4.3.1. Protection Status Relating to Socio-Economic Considerations

In the context of the proposed consequence category risk framework, regions of particular socio-economic value (e.g., local areas of high cultural significance, high significance to traditional activities, with designated special status such as United Nations Education, Science, and Cultural Organization (UNESCO) sites, etc.) could be mapped as having a particular Protection (P) category designation. Such information would indicate areas of high, moderate, and normal status (Table 3).

Table 3. Socio-economic protection categories.

Protection Status	Assigned Value
High	P = 3
Moderate	P = 2
Normal	P = 1

These regions can be geographically defined and easily communicated to operators in an automated fashion, including any special operating considerations required of vessels in these areas. Such an approach will help streamline regulatory implementation and inform operational planning and decision-making. Figure 3 illustrates how geographically referenced maps can be used to communicate regions with socio-economic protection status.

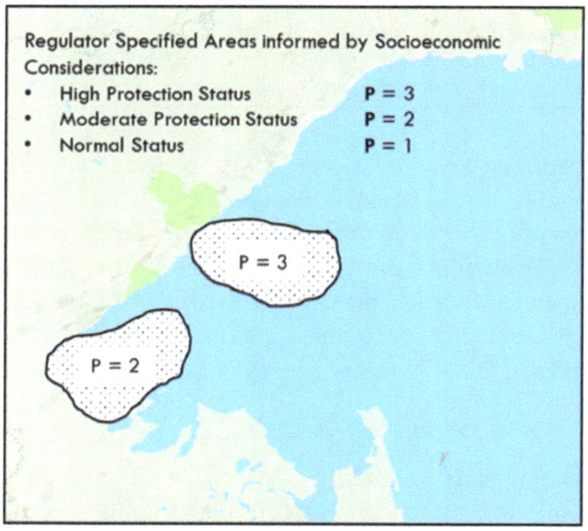

Figure 3. Examples of geographic regions defined as requiring socio-economic protection status.

4.3.2. Ecological Sensitivity Categories

Through ecological risk assessments and marine environmental assessments, it is possible to model and assess the sensitivity of different species and populations to identify the ecological sensitivity of geographic regions. Based on ecological characteristics (e.g., endangered species, nesting colonies, seasonal migrations, etc.), regions could be mapped as having ecological sensitivity (E) category designations that would indicate areas of high, moderate, and normal ecological sensitivity (Table 4). Policy information can be communicated in a similar fashion as proposed for regions requiring socio-economic protection status (Figure 3), including any special operating requirement for vessels in these regions.

Table 4. Ecological sensitivity categories.

Ecological Sensitivity Category	Assigned Value
High	E = 3
Moderate	E = 2
Normal	E = 1

4.3.3. Spill Consequence Categories

The spill risk for a given vessel will depend on the amount and type of potential contaminant carried onboard. Through oil spill risk assessments, vessels could be identified as having high, moderate, or normal levels of potential spill consequence (SC), as presented in Table 5. For example, chemical tankers carrying large volumes of hazardous liquids would be categorized as having a high SC value, while a smaller vessel with limited fuel (or that uses a more environmentally friendly fuel) may be assigned a lower SC value.

Table 5. Spill consequence categories for vessels.

Spill Consequence	Assigned Value
High	SC = 3
Moderate	SC = 2
Normal	SC = 1

As with other categories, regulators can specify what, if any, special operating considerations are required for higher spill consequence vessels.

4.3.4. Protection Status, Ecological Sensitivity, and Spill Consequence Index (PESCI)

The different consequence categories are combined to inform an operational exposure level. The process is referred to as the Protection Status, Ecological Sensitivity, and Spill Consequence Index (PESCI) method. A PESCI value is dependent on the socio-economic protection status (P) and ecological sensitivity (E) categories for the region, and the spill consequence category (SC) for the vessel. PESCI values are assigned in accordance with Table 6a,b,c, which corresponds to socio-economic protection category values (P) of 1, 2, and 3, respectively.

Table 6. PESCI values.

a. PESCI Values (P = 1)					b. PESCI Values (P = 2)					c. PESCI Values (P = 3)				
	3	2	3	4		**3**	3	4	5		3	4	5	6
SC	**2**	1	2	3	**SC**	**2**	2	3	4	**SC**	2	3	4	5
	1	0	1	2		**1**	1	2	3		1	2	3	4
		1	**2**	**3**			**1**	**2**	**3**			**1**	**2**	**3**
			E					**E**					**E**	

The PESCI value corresponds to an overall consequence category of high (C1), moderate (C2), or normal (C3), in accordance with Table 7.

Table 7. Consequence categories.

Consequence Category	PESCI Range
C1: high consequence	PESCI ≥ 3
C2: moderate consequence	1 < PESCI < 3
C3: normal consequence	PESCI ≤ 1

4.4. Operational Exposure Levels

The next step is determination of the operational exposure level, which is dependent on the life-safety and consequence categories, as detailed in Table 8. For example, ships with high life-safety (S1) are designated the highest operational exposure level (L1); ships with a moderate life-safety (S2) and a consequence category of high (C1), moderate (C2), or low (C3) are designated an operational exposure level of L1, L2, or L3, respectively.

Table 8. Operational exposure levels.

Life-safety Category				
	S1 (high)	L1	L1	L1
	S2 (moderate)	L1	L2	L3
	S3 (low)	L1	L3	L3
	S4 (unmanned)	L2	L4	L4
		C1 (high)	**C2** (moderate)	**C3** (low)
		Consequence Category		

4.5. RIV Adjustment for Exposure Levels

Finally, the proposed approach is incorporated into the existing POLARIS methodology through adjustment of the calculated RIVs, based on the determined operational exposure level. This is similar to the existing method to account for observed seasonal ice decay in POLARIS. RIV adjustment factors corresponding to operational exposure levels are presented in Table 9.

Table 9. Risk Index Value (RIV) adjustment factors for operational exposure levels.

Operational Exposure Level	RIV Adjustment Factor
L1	$RIV_{L1} = -2$
L2	$RIV_{L2} = -1$
L3	$RIV_{L3} = 0$
L4	$RIV_{L4} = +1$

Once the RIV adjustment factor is identified, a modified RIO is calculated following Equation (2).

$$RIO_{modified} = C1 \times (RIV1 + RIV_L) + C2 \times (RIV2 + RIV_L) + \cdots + Cn \times (RIVn + RIV_L) \quad (2)$$

where C1 ... Cn = concentration (in tenths) of each ice type within the ice regime, RIV1 ... RIVn = the corresponding standard RIVs for each ice type (following POLARIS); and RIV_L = the RIV adjustment factor.

The modified RIO is then used as the basis for the selection of one of three levels of operation, as per POLARIS: 'Normal', 'elevated operational risk', or 'operations subject to special consideration'. The overall process of the proposed risk assessment framework is presented in Figure 4.

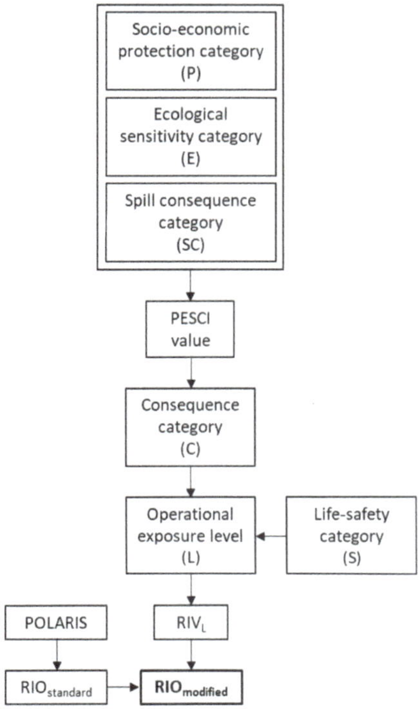

Figure 4. Risk assessment framework process.

5. Illustrative Example

To demonstrate the application of the proposed risk assessment framework and its impact on voyage planning and navigation, a fictitious waterway was considered (Figure 5).

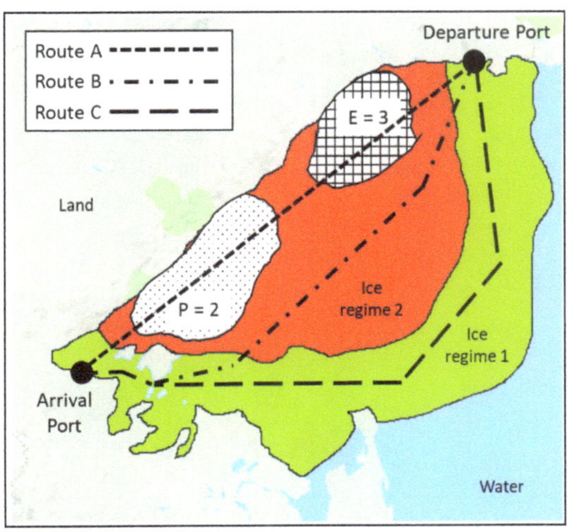

Figure 5. Waterway and routes for illustrative example.

Within the waterway, two regions were classified as having special ecological and socio-economic designations. The crosshatched region to the north was assigned a high ecological sensitivity category

(E = 3). To the south, the stippled region was assigned a moderate socio-economic protection status category (P = 2). Outside the regions of heightened environmental sensitivity, the ecological sensitivity and socio-economic protection category values were low (E = P = 1).

The operational exposure for five different vessels was assessed along three different routes. Based on the assessed risk level and resulting operational criteria, the optimal route for each vessel was identified.

5.1. Route Identification

The planned voyage departed from a northern port and arrived at a port in the south, as depicted in Figure 5. Two different ice regimes were present. The ice types and associated concentrations for each ice regime are presented in Table 10.

Table 10. Ice regimes for illustrative example.

Ice Regime	Ice Type	Concentration
Ice Regime 1	Thin First Year	5/10th
	Grey Ice	4/10th
	Open Water	1/10th
Ice Regime 2	Thick First Year	8/10th
	Thin First Year	1/10th
	Open Water	1/10th

Three different route options were available to transit from the departure port to the arrival port. Route A was the shortest distance, transiting along the coast through the more severe ice conditions (ice regime 2) and through the regions with heightened environmental sensitivity. Route B transited farther from the coast to avoid the regions of heightened sensitivity but remained in the more severe ice conditions. Route B was a longer distance than Route A. Route C was the longest distance, transiting the farthest from the coast to remain in the less severe ice conditions (ice regime 1).

Any vessel navigating through this waterway must acknowledge the ice regimes and the regions of heightened ecological and socio-economic sensitivity. The proposed risk assessment framework accounted for this and guided the decision on operating criteria for a given vessel.

5.2. Vessel Scenarios

To demonstrate the impact of the life-safety category and the spill consequence category, five different vessels were selected: A bulk carrier, an oil tanker, a cruise ship, a fishing vessel, and an autonomous ship. The vessel ice class, POB, associated life-safety category, and spill consequence value for each vessel is provided in Table 11.

Table 11. Vessel details for illustrative example.

Vessel Type	Ice Class	POB	Life-Safety Category	Spill Consequence Category
bulk carrier	PC5	55	S2	SC = 2
oil tanker	PC5	75	S2	SC = 3
cruise ship	PC5	2500	S1	SC = 1
fishing vessel	PC5	12	S3	SC = 1
autonomous	PC5	0	S4	SC = 1

The bulk carrier, with a crew of 55, had a moderate life-safety category (S2) and a moderate spill consequence (SC = 2). The oil tanker had a high spill consequence category (SC = 3) and a moderate life-safety category (S2). The cruise ship, with the highest number of passengers at 2500, received a high life-safety category (S1), but a low spill consequence (SC = 1). The fishing vessel had both a low life-safety category (S3) and a low spill consequence (SC = 1). The autonomous ship had an unmanned life-safety category (S4) and a low spill consequence (SC = 1). All vessels had an assigned ice class of PC5.

5.3. Results

In this example, four regions required separate risk assessments to evaluate and compare the routes. The four regions were ice regime 1, ice regime 2, the high ecologically sensitive region, and the moderate socio-economic protected status region. Three possible routes were considered here:

- Route A passed through all four regions and required assessment of each region.
- Route B avoided the regions of heightened environmental sensitivity and needed only to be assessed for both ice regimes.
- Route C passed only through ice regime 1.

The risk assessment results for each vessel in each of the four regions are presented in Table 12.

Table 12. Risk assessment results for the illustrative example.

Region	Vessel Type	PESCI	C	L	RIV_L	RIO (std)	RIO (mod)
Ice Regime 1 E = 1, P = 1	bulk carrier	1	C3	L3	0	25	25
	oil tanker	2	C2	L2	−1	25	15
	cruise ship	0	C3	L1	−2	25	5
	fishing vessel	0	C3	L3	0	25	25
	autonomous	0	C3	L4	1	25	35
Ice Regime 2 E = 1, P = 1	bulk carrier	1	C3	L3	0	5	5
	oil tanker	2	C2	L2	−1	5	−5
	cruise ship	0	C3	L1	−2	5	−15
	fishing vessel	0	C3	L3	0	5	5
	autonomous	0	C3	L4	1	5	15
High Ecological Sensitivity E = 3, P = 1	bulk carrier	3	C1	L1	−2	5	−15
	oil tanker	4	C1	L1	−2	5	−15
	cruise ship	2	C2	L1	−2	5	−15
	fishing vessel	2	C2	L3	0	5	5
	autonomous	2	C2	L4	1	5	15
Moderate Socio-economic Sensitivity E = 1, P = 2	bulk carrier	2	C2	L2	−1	5	−5
	oil tanker	3	C1	L1	−2	5	−15
	cruise ship	1	C3	L1	−2	5	−15
	fishing vessel	1	C3	L3	0	5	5
	autonomous	1	C3	L4	1	5	15

The standard RIOs (based on the current POLARIS methodology) were equivalent because the vessels had equivalent ice class (PC5). In ice regime 1 the standard RIO was 25 and in ice regime 2 the standard RIO was 5. This corresponded to an operational criterion of 'normal operations' in both ice regimes. The presence of regions with heightened environmental sensitivity did not impact the standard RIO.

The proposed risk assessment framework was applied to each vessel as described below.

5.3.1. Bulk Carrier

In ice regime 1, the bulk carrier received an overall consequence category C3 (normal) based on its moderate spill consequence value of 2. Combined with a moderate life-safety category S2, the bulk carrier was assigned an operational exposure level of L3 corresponding to an RIV adjustment factor of 0. The modified RIO was equivalent to the standard RIO. A similar result was observed for the bulk carrier in ice regime 2. The bulk carrier required no operational restrictions in ice regimes 1 or 2.

Due to the bulk carrier's moderate spill consequence value, the vessel received a modified RIO of −15 ('operation subject to special consideration') in the region of high ecological sensitivity. The bulk carrier should avoid operating in this region. In the region of moderate socio-economic sensitivity, the bulk carrier received a modified RIO of −5 ('elevated operational risk'). The bulk carrier may operate in this region with reduced speed, additional watching, or icebreaker escort.

The bulk carrier should avoid operating in the region of high ecological sensitivity due to the high operational exposure level. Route A was not an option. The bulk carrier can operate along Route B or C without any operational restrictions. Route B was the optimal choice as it is the shorter distance.

5.3.2. Oil Tanker

In ice regime 1, the oil tanker received a modified RIO of 15 ('normal operations'). In the more severe ice conditions of ice regime 2, the modified RIO was reduced to −5 ('elevated operational risk'). In both regions of heightened environmental sensitivity, the high spill consequence value of the oil tanker resulted in a modified RIO of −15 ('operation subject to special consideration').

The oil tanker should avoid Route A since operating in either of the environmentally sensitive regions imposes the strictest operational criteria. Navigation of the oil tanker along Route B would require reduced speeds, additional watch keeping, or icebreaker escort. Along Route C it could maintain 'normal operation' as this route had the lowest operational exposure.

Note that in the socio-economic protection status region, the moderate spill consequence category of the bulk carrier resulted in less restrictive operational criteria than the oil tanker (high spill consequence). This exemplified the impact of differences in spill consequence category on operating criteria.

5.3.3. Cruise Ship

In ice regime 1, the cruise ship received a modified RIO of 5 ('normal operations'). In ice regime 2 the cruise ship received a modified RIO of −15 ('operations subject to special consideration'), reflecting the severe life-safety consequences should an incident occur in this ice regime with 2500 passengers onboard. Due to the cruise ship's low spill consequence value, there was no additional consequence severity for operating in an environmentally sensitive zone. However, the modified RIO remained at −15 due to its high life-safety category in ice regime 2. Route C was the only option that allowed for 'normal operations' for a cruise ship of this size.

It is noted that the operational restrictions for the cruise ship would be much less severe if it were a smaller expedition cruise vessel with fewer than 500 people on board and/or if the vessel were built to a higher ice class.

5.3.4. Fishing Vessel

Due to the smaller numbers of POB, the fishing vessel fell within the S3 life-safety category. It is important to note that this designation was intended only to reflect the reduced scale of search and rescue response operations required to assist in an emergency for a vessel of this size compared to vessels with very large numbers of POB. Regardless of the life-safety category designation, adequate resources need to be in place to ensure safe operations and timely emergency response in *all* situations. This categorization should in no way be misinterpreted as placing different valuations on life-safety under different conditions. The correct interpretation here is that less emergency response resources are required to ensure a safe response for vessels with smaller numbers of POB than would be required to respond to vessels with larger numbers of POB.

Similarly, the low volumes of contaminants on board a smaller vessel, such as a fishing vessel, places it in a low spill consequence category. The goal in *all* cases is to prevent any potential environmental damage and minimize environmental impact. The designations proposed here reflect the fact that fewer resources would be required to respond to a potential environmental event and less ecological consequence would be expected for vessels that have lower spill consequence values.

For the fishing vessel, given its S3 and C3 designations, its operational exposure was assessed as low and its modified RIO was equivalent to the standard RIO. In the regions of ecological and socio-economic sensitivity, the fishing vessel received no adjustment to its operational criteria because its consequence severity was low. Normal operations can be maintained along any route. Route A was the optimal choice as it is the shortest distance.

5.3.5. Autonomous Vessel

The autonomous ship had an unmanned life-safety category and low spill consequence value. It received an RIV adjustment of +1. This increased the modified RIOs to 35 and 15 in ice regimes 1 and 2, respectively. The autonomous vessel can maintain normal operations along any route. Route A was the optimal choice as it was the shortest distance.

5.3.6. Comparison of Different Vessel Types

Having assessed all four regions in the waterway, we saw that regions of heightened environmental sensitivity only influence operational guidance to vessels with potentially higher spill consequence. This was, in turn, reflected in the viable route options available to each vessel type, as summarized in Table 13 below.

Table 13. Comparison of viable route options by vessel type.

Vessel Type	Route A	Route B	Route C
bulk carrier	'do not proceed'	'normal operations'	'normal operations'
oil tanker	'do not proceed'	'subject to special consideration'	'normal operations'
cruise ship	'do not proceed'	'do not proceed'	'normal operations'
fishing vessel	'normal operations'	'normal operations'	'normal operations'
autonomous vessel	'normal operations'	'normal operations'	'normal operations'

For the bulk carrier to maintain normal operations and not have to reduce its speed, it needed to avoid both the high ecologically sensitive region and the moderate socio-economic protected region. Route A was not an option.

The oil tanker should remain outside both environmentally sensitive regions. Route A was not an option. Route B was viable but required reduced speeds or other risk mitigation measures. Route C allowed for normal operations.

The large cruise ship was subjected to the most restricted operations as a result of its high life-safety category. To maintain normal operations, the cruise ship must select Route C. For cruise companies looking to build new vessels for operating in such regions, this information could play an important role in informing the selection of ice class and sizing new vessels, since smaller, higher ice class cruise ships would have greater operational range with fewer operability restrictions and lower costs for escort icebreakers.

The fishing vessel and autonomous vessel, given their lower life-safety and consequence categories, could proceed under normal operations along either Route 1 or Route 2. The autonomous vessel posed very low life-safety and environmental risk and was permitted to go into more severe ice conditions with less restrictions, should the owners wish to do so.

6. Benchmark Case Study

The example scenarios presented above are intended to illustrate the overall application of the framework. A benchmark case study is presented here to demonstrate the application of the proposed framework using inputs from Arctic marine shipping assessments. The case study considers a cruise vessel and an oil tanker navigating along the North West coast of Svalbard, Norway, during the summer season. Data to support the assignment of life-safety and environmental consequence categories were obtained from published Arctic marine risk assessments and environmental impact assessments.

Marchenko et al. [22] evaluated the life-safety risk for shipping in five different regions of the Arctic. Their evaluation was based on accident data, trends in ship activity, and expert knowledge elicitation. Consideration was given to regional dependencies, such as vessel traffic levels, environmental conditions, and private/government emergency response capacities. Risk matrices were developed for the five regions showing the frequency of different accident types for different vessels, and the severity of the consequences to human health.

Marchenko et al. [22] considered several accident types. For the purpose of this case study, focus was on damage by collision, recognizing that this captured collisions with ice as well as collisions with other ships or marine infrastructure. In the waters around Svalbard, cruise vessels are assessed as presenting a high life-safety risk (S1) and oil tankers are assessed as moderate (S2).

A relative spill consequence category for each vessel was assigned based on data from accidents with similar vessel types. The Exxon Valdez oil tanker spilled approximately 41,000 m^3 of crude oil after running aground off the coast of Alaska in 1989 [36]. The Motor Vessel (MV) Explorer cruise ship had approximately 210 m^3 onboard when it sank after striking an iceberg off the coast of Antarctica [37]. Based on these values, the oil tanker and the cruise ship were assigned relative spill consequence categories of high (SC = 3) and low (SC = 1), respectively.

The vessel-specific and environmental-specific consequence category values are presented in Table 14. For the purpose of this study, both vessels were assumed to have a Polar ice class of PC5.

Table 14. Vessel details for illustrative example.

Vessel Type	Ice Class	Life-Safety Category	Spill Consequence Category
cruise vessel	PC5	S1	SC = 1
oil tanker	PC5	S2	SC = 3

The Arctic Council [24] has reported areas of heightened ecological sensitivity throughout the Arctic, including around Svalbard. Their evaluation considers the impact of Arctic oil spills and other Arctic shipping-related threats on fish, bird, and mammal activities (e.g. migration, breeding, feeding, etc.). The regional sensitivities have a seasonal dependence. During the summer season, a large seabird breeding colony is present off the NW coast of Svalbard (Figure 6). This region is evaluated as having a high ecological sensitivity to oil spills (E = 3).

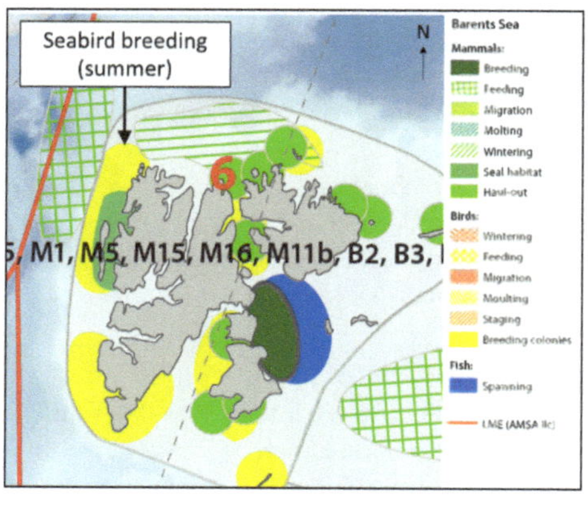

Figure 6. Areas of heightened ecological significance around Svalbard, Norway, (modified from Figure A.6 [24]).

The Arctic Council reported that the information necessary to evaluate culturally significant regions in the Arctic was not available at this time [24]. In the absence of this data, a default socio-economic protection category of low (P = 1) will be used for the purpose of this case study.

The ice conditions on 16 June 2019 were used for the case study, as reported by the Danish Meteorological Institute and presented in Figure 7. Two separate ice regimes overlap with the seabird breeding colony off the NW coast. The ice regimes are reported using Egg codes [38]. To the south and nearest to shore (Egg Code 'G') is one-tenth total concentration of thick first-year ice. Adjacent and to the north (Egg Code 'C') is more severe ice, reported four-tenths old ice, two-tenths thick first-year, and one-tenth medium first-year ice.

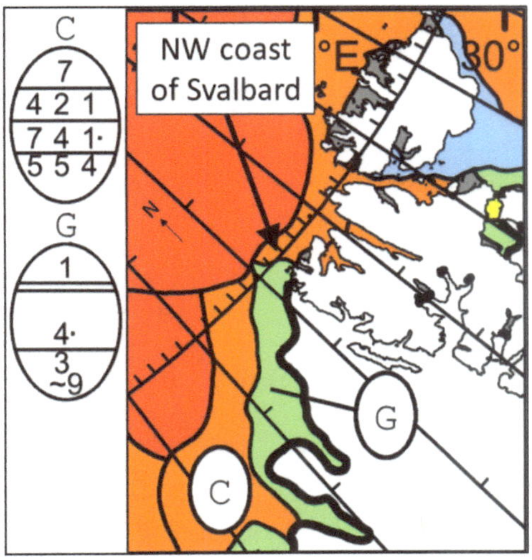

Figure 7. Ice conditions around the North West coast of Svalbard, Norway, 16 June 2019 (modified from Greenland Ice Chart, Danish Meteorological Institute, 16 June 2019 [39]).

The results of the risk assessment are presented in Table 15. Having been assigned a Polar class PC5, both vessels received standard RIOs of 27 in the ice regime 'G' (less severe ice conditions), and 2 in regime 'C' (more severe ice conditions). Based on the standard RIOs, the current POLARIS methodology would allow both vessels to undertake 'normal operations' in either ice regime.

Table 15. Risk assessment results for the benchmark case study.

Region	Vessel Type	PESCI	C	L	RIV_L	RIO (std)	RIO (mod)
Ice Regime G	oil tanker	1	C2	L1	−2	27	7
	cruise ship	3	C3	L1	−2	27	7
Ice Regime C	oil tanker	1	C2	L1	−2	2	−18
	cruise ship	3	C3	L1	−2	2	−18

Following the proposed risk assessment framework, both vessels were assessed as having a high operational exposure level, given the combinations of vessel type, ice conditions, and ecological sensitivity.

The cruise vessel had a high life-safety consequence value, which resulted in a high operational exposure level. In the less severe ice regime ('G'), the modified RIO was reduced to 2. This still allowed for 'normal operations' in this ice regime. In the more severe ice regime ('C'), the cruise vessel received a modified RIO of −18 ('operations subject to special consideration'). Given the life-safety consequence of cruise vessels operating around Svalbard, this ice regime should be avoided and the vessel should either choose an alternate route or delay operations in this region until ice conditions become less severe.

The oil tanker had a moderate life-safety consequence value and a high spill consequence value Combined with the high ecological sensitivity of the region, it received a high operational exposure level. In the less severe ice regime ('G'), the oil tanker received a modified RIO of 2, allowing 'normal operations'. In the more severe ice regime ('C'), the oil tanker received a modified RIO of −18, requiring 'operations subject to special consideration'. Given the severity of the ice conditions, the high spill consequence, and the high ecological sensitivity related to the presence of the summer seabird breeding colony, the oil tanker should avoid operations in this ice regime and select an alternate route.

7. Discussion

It has been demonstrated that increased Arctic shipping activity poses potential risk to life and environmental safety. Nevertheless, it must be recognized that Arctic shipping brings positive economic impacts to Arctic communities and nations [8]. A balance must be sought between the mitigation of risk and the realization of benefits [9].

The proposed framework does not aim to apply restrictions that are so stringent that the benefits of Arctic shipping cannot be realized. Under normal circumstances when vessels are transiting areas that do not have protected status and have normal ecological conditions (e.g., no sensitivities) then no changes in operating limits are required. If an area has been identified as having higher sensitivities or the vessel is carrying a large amount of potential contaminant or a large POB, adjustments to operating limits or deviation of route may be required.

The current methodology for assessing the operational limits of a vessel in ice (i.e., POLARIS) accounts only for the likelihood of ice damage. The risk assessment framework proposed here links with the current POLARIS methodology and provides operational guidance considering life-safety, environmental, and socio-economic consequences that can result from ice damage. Such an approach promotes Arctic marine safety and the protection of the Arctic environment and its stakeholders.

The proposed framework provides a methodology to incorporate varying risk perspectives into an operational decision support tool. It provides an avenue to capture risk profiles for different vessels, input from local communities and stakeholders, input from marine environmental impact assessments, and input from other marine risk assessments.

There are existing geographic information system (GIS)-based technologies that could support the calculation and communication of this information on the bridge of a ship, such as the Canadian Arctic Shipping Risk Indexing System (CASRAS), an e-navigation tool combining information on historic ice conditions, marine protected areas, community services, and mariner knowledge [40].

Marine operations, particularly in the Polar regions, are complex socio-technical systems. Risk management should take a holistic, multidisciplined approach. To move towards a more holistic assessment of risk, stakeholder engagement is necessary to establish the range of risk perspectives from those affected by and involved in Arctic shipping. The severity of consequences resulting from an accident in the Arctic will depend, in part, on availability and capacity of emergency response resources. Methods for incorporating system thinking is an area requiring future research.

There are ice class design attributes that contribute to mitigating life-safety and environmental risk (e.g., double hull oil tankers) and not all ice damage results in EER or an oil spill. These factors require consideration in the determination of a vessel's operational exposure level.

This framework provides a potential avenue for linking diverse research across a range of fields including engineering, as well as biological, physical, and social sciences, as demonstrated in the benchmark case study. A multidisciplinary approach can help inform decision making at operational and regulatory levels.

It is important to note that the specific values and levels of granularity proposed here for the various risk indices are starting points for discussion, to be debated and subjected to robust calibration exercises. This will require input from ongoing scientific research across multiple disciplines.

Next steps include further investigation of the proposed life-safety and consequence categories and operational exposure levels, as well as calibration of RIV adjustment factors. The efficacy of operational risk mitigation strategies requires further research and validation. Empirical data should be collected to strengthen the knowledge underlying the calibration of the proposed framework. Implementation of this methodology in a GIS-based software would simplify application of this approach and could accelerate verification and calibration.

8. Conclusions

A new framework was presented which augments the current POLARIS methodology to model the potential consequences of ice-induced damage. The framework incorporates the magnitude of life-safety and environmental consequences to support operational decision making for ships operating in Polar regions. The proposed framework complements the existing POLARIS methodology and guides the formulation of RIVs for varying risks and consequences. The outcome is that vessels of a given ice class with higher potential life-safety, environmental, and socio-economic consequences should be operated more conservatively. Mitigating measures, such as reducing the number of people on board, selecting more environmentally friendly fuels, specifying a higher ice class during design, or incorporating operational measures (e.g., support icebreakers) can enhance the operability of the vessel.

Author Contributions: Conceptualization, R.T. and T.B.; methodology, R.T. and T.B.; formal analysis, T.B.; investigation, T.B.; writing—original draft preparation, T.B.; writing—review and editing, R.T., B.V., P.K., F.K., and D.S.; visualization, T.B.; supervision, B.V., R.T., F.K., and D.S.; project administration, P.K. and B.V.; funding acquisition, P.K. All authors have read and agreed to the published version of the manuscript.

Funding: This research was funded by Lloyd's Register Foundation, grant number GA\100077, grant title: Recommended practise of scenario-based risk management for Polar waters.

Acknowledgments: The financial support of the Lloyd's Register Foundation is acknowledged with gratitude. Lloyd's Register Foundation helps to protect life and property by supporting engineering-related education, public engagement, and the application of research.

Conflicts of Interest: The authors declare no conflict of interest. The funders had no role in the design of the study; in the collection, analyses, or interpretation of data; in the writing of the manuscript, or in the decision to publish the results.

References

1. Kujala, P.; Goerlandt, F.; Way, B.; Smith, D.; Yang, M.; Khan, F.; Veitch, B. Review of risk-based design for ice-class ships. *Mar. Struct.* **2019**, *63*, 181–195. [CrossRef]
2. Smith, D.; Veitch, B.; Khan, F.; Taylor, R. Using the FRAM to understand Arctic ship navigation: assessing work processes during the Exxon Valdez grounding. *Int. J. Mar. Navig. Saf. Sea Transp.* **2018**, *12*, 447–457. [CrossRef]
3. Bergström, M.; Hirdaris, S.; Valdez Banda, O.A.; Kujala, P.; Thomas, G.; Choy, K.L.; Stefenson, P.; Nordby, K.; Li, Z.; Ringsberg, J.W.; et al. Towards holistic performance-based conceptual design of Arctic cargo ships. In Proceedings of the 13th Intern. Marine Design Conference, Helsinki, Finland, 10–14 June 2018; pp. 831–839.
4. Haimelin, R.; Goerlandt, F.; Kujala, P.; Veitch, B. Implications of novel risk perspectives for ice management operations. *Cold Reg. Sci. Technol.* **2017**, *133*, 82–93. [CrossRef]
5. IMO. International code for ships operating in polar waters (Polar Code). MSC 94/21/Add. 1 Annex 6. International Maritime Organization. 2014. Available online: https://edocs.imo.org/FinalDocuments/English/MEPC68-21-ADD.1(E).doc (accessed on 3 June 2019).
6. IMO. *Guidance on Methodologies for Assessing Operational Capabilities and Limitations in Ice*; MSC.1/Circ.1519; International Maritime Organization (IMO): London, UK, 2016; Available online: https://edocs.imo.org/FinalDocuments/English/MSC.1-CIRC.1519(E).docx (accessed on 4 July 2019).
7. Basaran, I. The future of Arctic navigation: cooperation between the International Maritime Organization and Arctic Council. *J. Marit. Law Commer.* **2017**, *48*, 35–52.
8. *Arctic Marine Shipping Assessment 2009 Report*, 2nd ed.; Arctic Council's Protection of the Arctic Marine Environment (PAME): Akureyri, Iceland, 2009.

9. Hauser, D.D.W.; Laidre, K.L.; Stern, H. Vulnerability of Arctic marine mammals to vessel traffic in the increasingly ice-free Northwest Passage and Northern Sea Route. *Proc. Natl. Acad. Sci. USA* **2018**, *115*, 7617–7622. [CrossRef] [PubMed]
10. Santos, C.F.; Carvalho, R.; Andrade, F. Quantitative assessment of the differential coastal vulnerability associated to oil spills. *J. Coast. Conserv.* **2013**, *17*, 25–36. [CrossRef]
11. *Petroleum and Natural Gas Industries – Arctic Offshore Structures*; ISO/FDIS 19906:2010(E); International Organization for Standardization (ISO): Geneva, Switzerland, 2010.
12. Kujala, P.; Kämäräinen, J.; Suominen, M. Validation of the new risk based design approaches (POLARIS) for Arctic and Antarctic operations. In Proceedings of the International Conference on Port and Ocean Engineering under Arctic Conditions, Delft, The Netherlands, 9–13 June 2019.
13. Taylor, R.S.; Richard, M.; Hossain, R. A Probabilistic High-Pressure Zone Model for Local and Global Loads During Ice-Structure Interactions. *J. Offshore Mech. Arct. Eng.* **2019**, *141*, 051604. [CrossRef]
14. Freeman, R.E. Design of ships and offshore structures: A probabilistic approach for multi-year ice and iceberg impact loads for decision-making with uncertainty. Ph.D. Thesis, Memorial University of Newfoundland, St. John's, NL, Canada, 2016. Available online: http://research.library.mun.ca/id/eprint/12411 (accessed on 1 December 2019).
15. Taylor, R.S.; Richard, M. Development of a probabilistic ice load model based on empirical descriptions of high pressure zone attributes. In Proceedings of the International Conference on Ocean, Offshore and Arctic Engineering (OMAE), San Francisco, CA, USA, 8–13 June 2014.
16. Jordaan, I.J. Mechanics of ice–structure interaction. *Eng. Fract. Mech.* **2001**, *68*, 1923–1960. [CrossRef]
17. *SOLAS: International Convention for the Safety of Life at Sea, 1974: Resolutions of the 1997 SOLAS Conference Relating to Bulk Carrier Safety*; IMO: London, UK, 1998.
18. *MARPOL Consolidated Edition 2011: Articles, Protocols, Annexes and Unified Interpretations of the International Convention for the Prevention of Pollution from Ships, 1973, as Modified by the 1978 and 1997 Protocols*, 5th ed.; IMO: London, UK, 2011.
19. *Revised Guidelines for Formal Safety Assessment (FSA) for Use in the IMO Rule-Making Process*; MSC-MEPC.2/Circ.12/Rev.2; IMO: London, UK, 2018; Available online: http://www.imo.org/en/OurWork/Safety/SafetyTopics/Documents/MSC-MEPC%202-Circ%2012-Rev%202.pdf (accessed on 4 July 2019).
20. Fedi, L.; Etienne, L.; Olivier, F.; Rigot-Müller, P.; Stephenson, S.; Cheaitou, A. Arctic navigation: stakes, benefits and limits of the POLARIS system. *J. Ocean Technol.* **2018**, *13*, 54–67.
21. Bond, J.; Hindley, R.; Kendrick, A.; Kämäräinen, J.; Kuulila, L. Evaluating risk and determining operational limitations for ships in ice. In Proceedings of the Arctic Technology Conference, Houston, TX, USA, 5–7 November 2018.
22. Marchenko, N.A.; Andreassen, N.; Borch, O.J.; Kuznetsova, S.Y.; Ingimundarson, V.; Jakobsen, U. Arctic shipping and risks: emergency categories and response capacities. *Int. J. Mar. Navig. Saf. Sea Transp.* **2018**, *12*, 107–114. [CrossRef]
23. Helle, I.; Jolma, A.; Venesjärvi, R. Species and habitats in danger: estimating the relative risk posed by oil spills in the northern Baltic Sea. *Ecosphere* **2016**, *7*, 1–17. [CrossRef]
24. Identification of Arctic marine areas of heightened ecological and cultural significance: Arctic Marine Shipping Assessment (AMSA) IIc. Arctic Monitoring and Assessment Programme (AMAP), AMAP/CAFF/SDWG, Oslo, Norway. 2013. Available online: https://www.amap.no/documents/doc/identification-of-arctic-marine-areas-of-heightened-ecological-and-cultural-significance-arctic-marine-shipping-assessment-amsa-iic/869 (accessed on 26 September 2019).
25. Nevalainen, M.; Helle, I.; Vanhatalo, J. Preparing for the unprecedented – towards quantitative oil risk assessment in the Arctic marine areas. *Mar. Pollut. Bull.* **2017**, *114*, 90–101. [CrossRef] [PubMed]
26. Dawson, J.; Johnston, M.E.; Stewart, E.J. Governance of Arctic expedition cruise ships in a time of rapid environmental and economic change. *Ocean Coast. Manag.* **2014**, *89*, 88–99. [CrossRef]
27. Simões Ré, A.; Veitch, B. Simões Ré; A.; Veitch, B. Escape-evacuation-rescue response in ice-covered regions. In Proceedings of the International Offshore and Polar Engineering Conference (ISOPE 2008), Vancouver, BC, Canada, 6–11 July 2008.
28. Aven, T.; Anderson, H.B.; Cox, T.; Droguett, E.L.; Greenberg, M.; Guikema, S.; Zio, E. Risk analysis foundations. Society for Risk Analysis 2015. Available online: https://pdfs.semanticscholar.org/b643/dbc0c69946ce67c74172b14ad1b6a054440a.pdf (accessed on 22 May 2019).

29. Goerlandt, F.; Montewka, J. Maritime transportation risk analysis: review and analysis in light of some foundational issues. *Reliab. Eng. Syst. Saf.* **2015**, *138*, 115–134. [CrossRef]
30. Parviainen, T.; Lehikoinen, A.; Kuikka, S.; Haapasaari, P. Risk frames and multiple ways of knowing: coping with ambiguity in oil spill risk governance in the Norwegian Barents Sea. *Environ. Sci. Policy* **2019**, *98*, 95–111. [CrossRef]
31. Naseri, M.; Barabadi, A. On context, issues, and pitfalls of expert judgement process in risk assessment of Arctic offshore installations and operations. In Proceedings of the IEEE International Conference on Industrial Engineering and Engineering Management (IEEM), Bangkok, Thailand, 16–19 December 2018.
32. Aven, T. Practical implications of the new risk perspectives. *Reliab. Eng. Syst. Saf.* **2013**, *115*, 136–145. [CrossRef]
33. Montewka, J.; Goerlandt, F.; Kujala, P.; Lensu, M. Towards probabilistic models for the prediction of a ship performance in dynamic ice. *Cold Reg. Sci. Technol.* **2015**, *112*, 14–28. [CrossRef]
34. Fu, S.; Zhang, D.; Montewka, J.; Yan, X.; Zio, E. Towards a probabilistic model for predicting ship besetting in ice in Arctic waters. *Reliab. Eng. Syst. Saf.* **2016**, *155*, 124–136. [CrossRef]
35. Bergström, M.; Ove Erikstad, S.; Ehlers, S. Assessment of the applicability of goal- and risk-based design on Arctic sea transport systems. *Ocean Eng.* **2016**, *128*, 183–198. [CrossRef]
36. *Marine Accident Report: Grounding of the US Tankship EXXON VALDEZ on Bligh Reef, Prince William Sound, near Valdez, Alaska, March 24, 1989*; NTSB/MAR-90/04; National Transportation Safety Board: Washington, DC, USA, 1990; Available online: https://www.arlis.org/docs/vol1/B/22590091.pdf (accessed on 1 March 2019).
37. *Report of Investigation in the Matter of Sinking of Passenger Vessel EXPLORER (O.N. 8495) 23 November 2007 in the Bransfield Strait Near the South Shetland Islands*; Bureau of Maritime Affairs: Monrovia, Liberia, 2009.
38. *MANICE Manual of Standard Procedures for Observing and Reporting Ice Conditions*, 9th ed.; Environment: Ottawa, ON, Canada, 2005.
39. *Greenland Ice Chart*; Danish Meteorological Institute: Copenhagen, Denmark, 2019.
40. Charlebois, L.; Kubat, I.; Lamontagne, P.; Burcher, R.; Watson, D. Navigating in polar waters with CASRAS. *J. Ocean Technol.* **2017**, *12*, 43–52.

© 2020 by the authors. Licensee MDPI, Basel, Switzerland. This article is an open access article distributed under the terms and conditions of the Creative Commons Attribution (CC BY) license (http://creativecommons.org/licenses/by/4.0/).

applied sciences

MDPI

Article

Numerical Simulation of Ship Oil Spill in Arctic Icy Waters

Wei Li [1,*], Xiao Liang [2], Jianguo Lin [1,*], Ping Guo [1], Qiang Ma [3], Zhenpeng Dong [2], Jiamin Liu [1], Zhenhe Song [1] and Hengqi Wang [1]

1 College of Environmental Science and Engineering, Dalian Maritime University, Dalian 116026, China; guoping@dlmu.edu.cn (P.G.); liujiamin201700@126.com (J.L.); songzhenhhhhh@126.com (Z.S.); wanghengqi20171231@126.com (H.W.)

2 College of Naval Architecture and Ocean Engineering, Dalian Maritime University, Dalian 116026, China; dongzhenpeng@126.com (Z.D.); xiaoliang@dlmu.edu.cn (X.L.)

3 Naval Architecture and Marine Engineering College, Shandong Jiaotong University, Weihai 264200, China; qiangma1970qq@126.com

* Correspondence: weiwei99231@dlmu.edu.cn (W.L.); ljglin@dlmu.edu.cn (J.L.)

Received: 14 January 2020; Accepted: 14 February 2020; Published: 19 February 2020

check for updates

Featured Application: The trajectory and geometry of oil film could be predicted at a certain future moment to provide an effective method for the emergency treatment of maritime oil spill.

Abstract: This paper presents a three-dimensional numerical simulation model of an oil spill for application in emergency treatment methods under icy water conditions. The combined effects of wind, wave, current and ice implemented in our model correspond to Arctic Ocean conditions. A discrete element method combined with an overset grid was adopted to track the trajectory movements of oil film with medium-density ice floes and simulate the flow field of moving ice of large displacement in six degrees of freedom (6DOF). The probability of oil spill area extensions were estimated by a response surface method (RSM). Results showed reduced risk of pollution in icy water conditions and greater drift action of oil film. Accordingly, the spraying location and quantity of oil-dispersant could be rapidly specified.

Keywords: arctic ice; oil spill; discrete element method; pollution prediction; 6DOF model

---

1. Introduction

In recent years, ice-free areas in the Arctic Ocean increased especially during summer months. This increased business utilizing deep waters in this region [1]. With the opening of the Arctic Shipping Route, oil spill accidents from irregular operations or ship collisions in icy Arctic waters can cause tremendous damage to the aquatic ecosystem [2]. The oil spill is a serious pollution problem for the marine environment, as well as a concern for the survival quality of fish and algae [3], particularly in fragile Arctic regions. The hydrocarbons contained in petroleum can seriously damage the physiological and biochemical processes of marine organisms. Therefore, it is imperative to study emergency methods to treat ship oil spills. The diffusion and drift of oil films on the water surface, as well as the spread of oil under ice or trapped by ice floes, can be simulated to protect the Arctic maritime environment.

Oil-spill behaviors in ice-free waters have been researched systematically since the 1970s. In this work, the convection diffusion equation was considered suitable for the prediction of oil-spill diffusion under water [4]. In this paper a buoyant jet model based on the Lagrangian control volume method was established. The method considered both the diffusion and dissolution processes of oil spills [5]. Additionally, the "particle flow" method was used to calculate the transport positions of oil particles at each time step to simulate the transport process of the oil film and associated oil-spill trajectory [6]. The trajectory of oil-spill diffusion on the sea surface was presented visually using a discrete phase model (DPM) in FLUENT software [7]. The Sea Track Web system was applied for oil drift forecasting [8].

The forecasting of oil spills in water is complex, and the presence of ice in cold waters increases further the difficulty of this research. This is because of the fundamental interaction among different ice blocks that possess different equilibrium thickness in cold and warm open waters [9]. Depending on the natural geographic condition of the Bohai Sea, a dynamic sea ice model has been developed to understand the effects of various forces on sea ice movement [10]. Based on continuum theory, the viscous-plastic constitutive relationship of sea ice was analyzed to forecast the movement of an oil spill with different ice densities [11]. The morphology of "oil particles" was used to establish the drift-diffusion model of the oil spill [12]. Afterwards, the evaporation and emulsification processes were added to the forecasting of oil spills in the pack ice area to correct the drift equation [13]. A surface current model and a sea ice model were combined to simulate the oil slick movement process, including drifting, spreading, evaporation and emulsification [14]. However, in its current model, the waves on the sea surface were neglected, and only the oil spill on the ice surface could be forecasted. The movement of an oil spill under ice was observed with the help of 3-D under-ice imagery from an autonomous underwater vehicle [15]. Nevertheless, the spread of an oil spill under the ice was difficult to predict precisely.

The objective of this work has been to consider synthetically the real environmental elements of wind, current as well as waves in the Icelandic Sea of the Arctic Ocean with the aim to simulate an authentic environment flow field for the prediction of oil-spill trajectory. To achieve this, the multi-phase volume of fluid method (VOF) [16] and discrete element methods (DEM) [17] were adopted to track the non-linear free surface and establish a three-dimensional icy waters model. The Eulerian–Lagrangian method was applied to simulate the movement of the oil spill in ice-infested and ice-free waters. To shorten the forecasting time, the response surface method (RSM) [18] was used to fit the functional relationship between the oil-spill variables and various responses.

2. Modelling Methods

Based on the unique characteristics of the Arctic sea, the combined influences of wind, current, wave and sea ice were considered with the aim to establish a three-dimensional numerical model of a ship oil spill. The oil, air and water were separated into three insoluble phases. The air phase assumed steady wind and the water phase assumed combination of current and non-linear waves. The VOF was used to track the free surface of the air–water phase interface and DEM helped idealize the interactions between ice floes. A Eulerian–Lagrangian discrete method was applied to simulate the movement of oil. Accordingly, instead of static analysis, a six degrees of freedom (6DOF) dynamic simulation code was introduced to control the pitching motion and horizontal displacement of ices. The oil trajectory dynamic variation was achieved by coupling the movement of floating ice and oil drift.

2.1. VOF Method

The sea surface with wave was presumed as an air–water interface, and the oil spill from underwater to water surface was a three-phase oil–water–air flow. Therefore, the VOF was adopted to track the free surface of the water [19]. The VOF method tracked the fluid flow in each control cell by constructing a fluid volume fraction function to inform the free surface shape with its function and derivative values [20]. When the specified phase fluid was located at the lower left of the grid element, the lower left vertex was the datum, while the fluid was in another orientation, and the datum was obtained by coordinate flipping. The outward normal in a free surface grid was taken as the unit of the outward normal vector. The horizontal direction was the X axis and the vertical direction was the Y axis. The fluid volume fraction function F_q satisfied Equation (1) [21]. If $F_q = 1$, the entire cell consisted of the q-th phase fluid; if $F_q = 0$, there was no q-th phase fluid in the cell; and if $0 < F_q < 1$, the cell was considered as an interface cell. F_q was a continuous function of time and space, and the sum of the volume ratios $\sum F_q$ of each phase q was 1. In the VOF method, the physical parameter φ was calculated by Equation (2).

$$\frac{\partial F_q}{\partial t} + \frac{\partial(uF_q)}{\partial x} + \frac{\partial(vF_q)}{\partial y} = 0, \sum F_q = 1 \tag{1}$$

$$\varphi = \sum \varphi_q F_q \tag{2}$$

2.2. Discrete Element Method (DEM)

The interactions among sea ice cells were modeled by DEM. In the normal direction, the cells in DEM can be regarded as being connected by a damper and a spring in parallel. In the tangential direction, they could be assumed connected by a spring, a damper and a sliding friction device [22]. During the interaction process, the viscoelastic forces among cells resulting from their relative velocities and elastic deformations were considered, and the shear force was computed according to Mohr–Coulomb [23] friction law. The normal force between two ice cells was given by Equation (3). The tangential contact force between oceanic ices was expressed by Equation (4), and K_t was set to 60% of normal stiffness.

$$F_n^n = K_n \delta_{ij} - C_n V_{ij} \cdot \mathbf{n} \tag{3}$$

$$F_t^n = \min\left(\left(F_t^{n-1} - K_t \Delta t \left(V_{ij} \cdot \mathbf{t}\right)\right), \mu F_n^n\right) \tag{4}$$

2.3. Eulerian–Lagrangian Method

The diffusion and drift of oil film were computed by the conservation law of mass and momentum based on the convection–diffusion equation to simplify the coupling of the particle motion, chemical dynamic and hydrodynamic equations [24].

The ocean transport convection–diffusion equation was expressed in Lagrangian format as Equations (5)–(7). Differential dispersion was expressed by Equations (5)–(8) [25].

$$\frac{DC}{Dt} = \frac{\partial}{\partial z}\left(k\frac{\partial C}{\partial z}\right) + K_H \nabla^2 C \tag{5}$$

$$\nabla^2 = \frac{\partial}{\partial x^2} + \frac{\partial}{\partial y^2} \tag{6}$$

$$\frac{D}{Dt} = \frac{\partial}{\partial t} + u\frac{\partial}{\partial x} + v\frac{\partial}{\partial y} + w\frac{\partial}{\partial z} \tag{7}$$

$$\frac{DC}{Dt} = \frac{C_Q^{K+1} - C_P^K}{\Delta t} \tag{8}$$

After the time Δt, this micelle arrived at position Q, and position P could be calculated by the velocity field along the trajectory of the substance micelle [26].

The amount of oil spill was expressed as a release of particles. Thus, when the thickness of an oil film on the water surface reached a terminal value (the mass balance of oil spill), the oil layer could be regarded as a spatial mapping distribution of particles. Slick thickness h in grid cells was calculated by Equation (9).

$$h = \sum \frac{V_i}{\Delta A} \tag{9}$$

The Euler–Lagrangian method was then used to establish a drift–diffusion model of oil spill on the seawater surface. The equation of motion was expressed as Equation (10). The diffusion coefficient of the oil film was expressed as Equation (11).

$$\frac{\partial h}{\partial t}+\frac{\partial h(u+\tau_x/f)}{\partial x}+\frac{\partial h\left(v+\tau_y/f\right)}{\partial y} = \frac{\partial^2 D_s h}{\partial x^2}+\frac{\partial^2 D_s h}{\partial y^2}+R_h \tag{10}$$

$$D_s = gh^2(\rho_w-\rho_o)\rho_o/\rho_w f \tag{11}$$

3. Numerical Simulation of Oil Spill

3.1. Simulation of Oil-Spill Drift

The wind and current were assumed to affect the centroid of the oil film. The Eulerian–Lagrangian tracking method was adopted to simulate the drift trajectory of the oil film centroid on the water surface as Equation (12) which explained the initial position S_0 of the oil film centroid drifting to the new position S after a time Δt. U_L was calculated by Equation (13) and a_w was commonly ranging from 0.03 to 0.04. S was expressed as Equations (14) and (15). The velocity vector of oil-spill drift is shown in Figure 1.

$$S = S_0+\int_0^{0+\Delta t} U_L dt \tag{12}$$

$$U_L = a_f U_f + a_w U_w \tag{13}$$

$$x' = x_0+u\Delta t+i_w{\cdot}U_w{\cdot}sin\theta \tag{14}$$

$$y' = y_0+v\Delta t+i_w{\cdot}U_w{\cdot}cos\theta \tag{15}$$

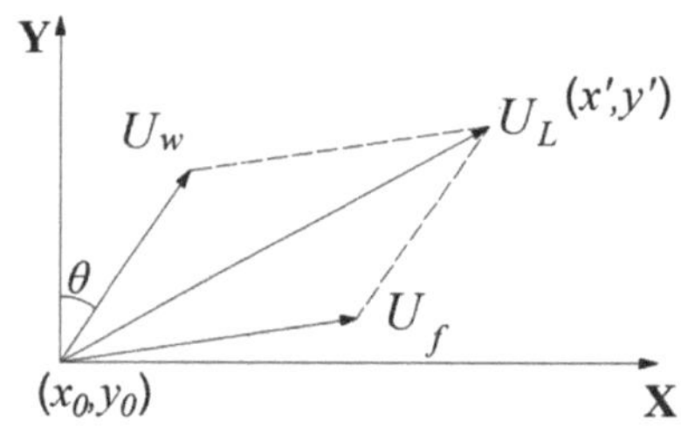

Figure 1. Velocity vector synthesis of oil-spill drift.

3.2. Simulation of Ice-Floes Drift

The short-time prediction model of ice-floes movement was based on momentum conservation and mass continuity equations, therein, the critical step was to determine the constitutive relation of the ice internal force. The constitutive relation of the ice floes could be divided into two categories, based on the continuum and discrete medium. According to different ice densities, the viscoplastic constitution was adopted for continuous ice floes, and the collision-type constitutive relation was employed for the edge ice belts with discrete medium properties. Two types of equations were combined for the hydrodynamic model of ice floes, as shown in Equations (16)–(18). Equation (16) presented the momentum balance equation. Equations (17) and (18) presented mass conservation; and *diffusion* represented the term of energy diffusion. In the prediction model, the Coriolis force, the ocean surface slope and thermodynamic terms were assumed negligible.

$$M\frac{D\vec{U}}{Dt} = -Mf\vec{k}\times\vec{U}+A\vec{\tau_a}+A\vec{\tau_w}-Mg{\cdot}gradH+\vec{F} \tag{16}$$

$$\frac{\partial h_a}{\partial t} + \vec{U} \cdot \nabla h_a = \Phi_h + diffusion \quad (17)$$

$$\frac{\partial A}{\partial t} + \vec{U} \cdot \nabla A = \Phi_A + diffusion \quad (18)$$

3.3. Numerical Model

A continuous oil-spill accident, which happened in the geographical range of 66°24′ N to 66°27′ N and 15°42′ W to 15°54′ W, was simulated here. It was assumed as a static point source in the environmental conditions of waters near Iceland. As shown in Figure 2, the location of the oil spill was identified as Point A (66.43° N, 15.79° W). The parameters used in this model are shown in Table 1. Some parameters on the behavior of oil refer to the obtained marine characteristics [27] and the sensitivity studies by Ross and Dickins [28].

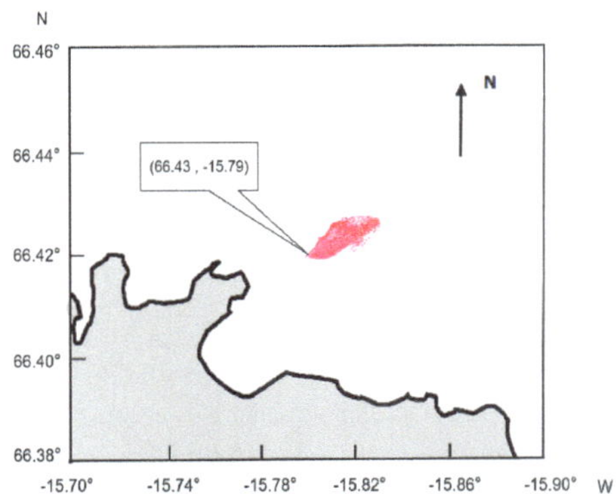

Figure 2. Computational domain of Icelandic waters.

Table 1. The parameters used in the oil-spill model.

Properties	Parameters
Location	66.43° N, 15.79° W
Release situation	Accidental ship leakage
Spill rate	15 kg/s
Spill duration	200 s
Oil density	840 kg/m^3
Oil viscosity	500 mPa·s
seawater density	1025 kg/m^3
ice density	910 kg/m^3
Ice diameter	5 m
Ice concentration	Medium (30–80%)

The computational domain (Figure 3) had a dimension of 2000 × 1000 × 500 m, with the air domain on the top, water domain on the bottom, and the free surface at the height of 400 m. The smallest grid cell was 0.005 m, and there were 1.35 million cells in total. The oil-spill spout was set as 2 m in diameter. The oil spurted vertically from 0.5 m underwater, and was located at 100 m to the left end of the domain. The left side of the domain was the speed inlet; the right side was the pressure outlet; the upper and lower boundaries were wave-eliminating.

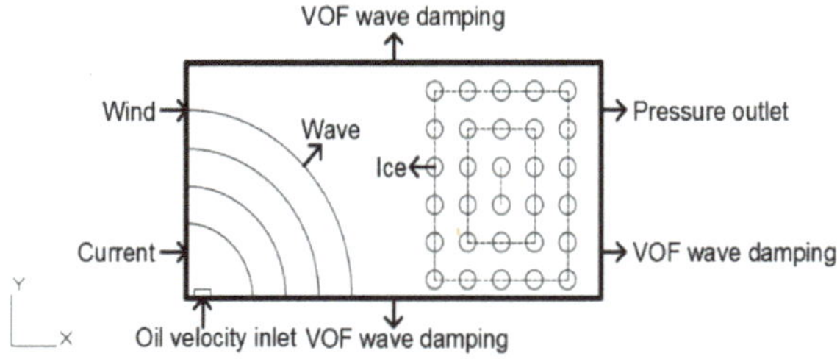

Figure 3. Computational domain.

The grids were locally refined around the free surface and the oil spout. The girds refinement aimed to capture the flowing details preferably without significantly increasing the amount of calculation, especially in the small domain near the free surface and the oil spout (to guarantee grids with more than 20 layers in the wave height direction, 80 grids at one wavelength, surface grids refinement near the oil spout of 25%), as shown in Figure 4, and a three-dimensional implicit unsteady model of the marine environment was established. The dominating equations involved in the numerical calculation were the fluid continuity equation, mass conservation equation, and momentum equation (Navier–Stokes equation) [29]. The VOF method was adopted to describe the three-phase oil–water–air flow. The waves were fifth-order VOF waves with a wavelength of 60 m and a wave height of 1.5 m, which were initialized and shown in Figure 5. The drift of ice floes was simulated by DEM. An ice ejector was constructed using Lagrangian discrete particles, and each ice floe was simulated as a round, flat particle with a thickness of 0.2 m [30]. It was presumed that there were mutual collisions between ice floes without fragmentation and the particles were randomly sprayed in accordance with the derivative components of grid arrangement. After the ejector generated ice floe particles for 100 s, the oil-spill accident was initiated. This time point could be approximated as the completion of the ice-field initialization. The $k-\varepsilon$ model was selected to present the turbulence mode and the computational step was set to 1 s.

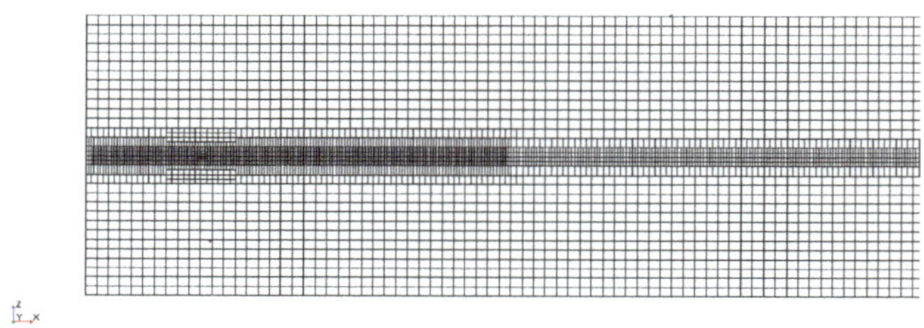

Figure 4. Grids' refinement around the oil spout.

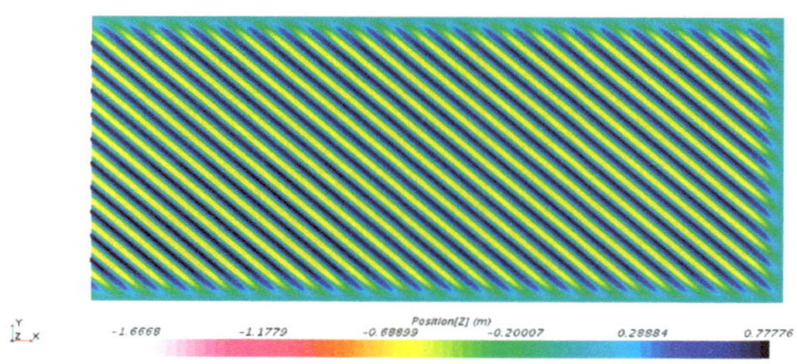

Figure 5. Initialization of free surface wave.

A series of meshes ranging from 10.2 to 14.6 million elements were made as an example case, where 10.2, 12.2 and 14.6 million elements are designated as the coarse mesh, the medium mesh and the fine mesh, respectively. The computational time of the medium mesh was approximately 80 h, which was 3 times faster than that of the fine mesh using the 64-bit operating computational system, 3.8 GHz, 32 Core Processors with 8 GB RAM. The computational time of the coarse mesh was approximately 50 h with the same computational parameters. Thus, the medium mesh parameters were considered appropriate for this study.

3.4. Characteristics of Oil Spill in Icy and Ice-Free Waters

The oil spill behavior was simulated for both open (ice-free) and icy water conditions under the same sea state. The oil diffusion trajectories for open waters and icy waters are shown in Figures 6 and 7, respectively. The blue color represents the oil film and the other colors represent the ice floes on the sea surface. In icy waters, the extent of the oil spill was obviously suppressed and its distribution was relatively narrow. Its maximum drift distance and maximum diffusion area radius were much smaller than those in open waters. At the time of 500 s, in open waters, the maximum drift distance of the oil spill was 1.2 km in the horizontal direction and 0.6 km in the vertical direction; in icy waters, the maximum oil-drift distance was only 0.7 km in the horizontal direction and 0.1 km in the vertical direction. The main direction of oil-spill diffusion was affected by wind and waves under both conditions. In ice-free waters, the oil film was widely spread on the sea surface, whereas in icy waters, a band-like distribution was observed.

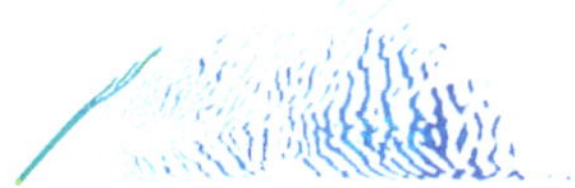

(a) T = 200 s

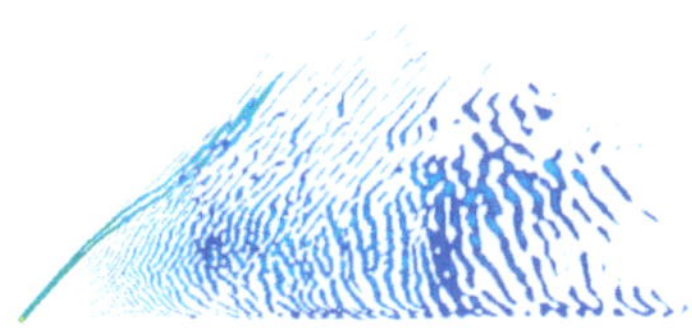

(b) T = 350 s

Figure 6. *Cont.*

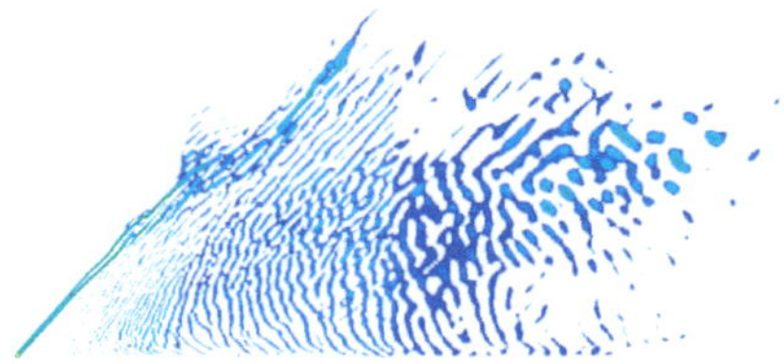

(**c**) T = 500 s

Figure 6. Oil diffusion trajectories for open waters.

(**a**) T = 200 s

(**b**) T = 350 s

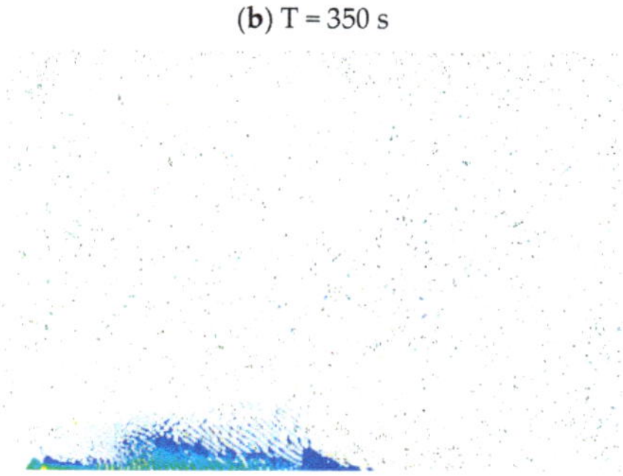

(**c**) T = 500 s

Figure 7. Oil diffusion trajectories for icy waters.

Although the presence of ice was observed slowing down the oil diffusion rate, the blending of ice and oil would increase the difficulty of oil recovery. The response measures for the oil spill should, therefore, be adopted in a timely manner according to the oil-spill trajectory. In the forecast area of the oil spill, the enclosure and control facilities should be deployed to prevent the wide-area oil-spill accidents.

In icy waters, the oil diffusion was affected by wave action on the free surface. These diffusion trajectories are shown in Figures 8 and 9, where the red portion represents the seawater, the blue portion represents air and the interface between the two is the free surface. The black parts indicated the ice floe particles that rose and fell with the waves. The oil diffusion was represented by contours and the oil spout was located at y = 0 m. Oil drift could be observed along the horizontal direction with time and the thickness of the oil film changed continuously, representing a simultaneous process of drift and diffusion.

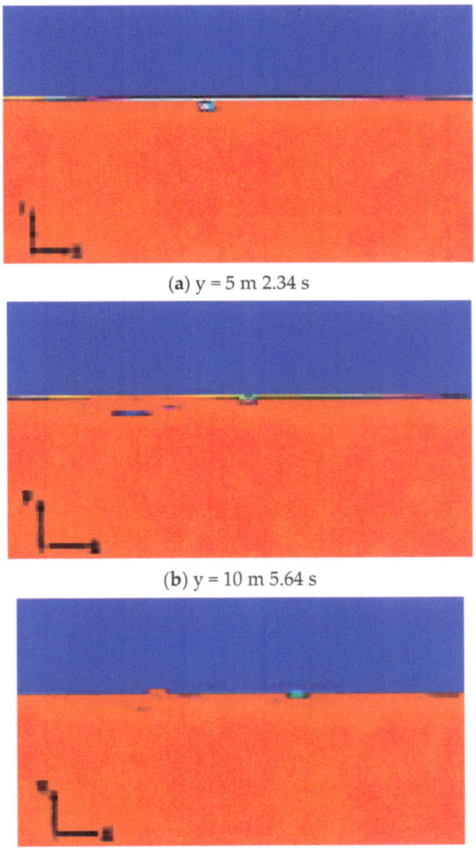

(**a**) y = 5 m 2.34 s

(**b**) y = 10 m 5.64 s

(**c**) y = 15 m 9.48 s

Figure 8. Three-phase distribution of oil–water–air for icy waters at the oil emerging time.

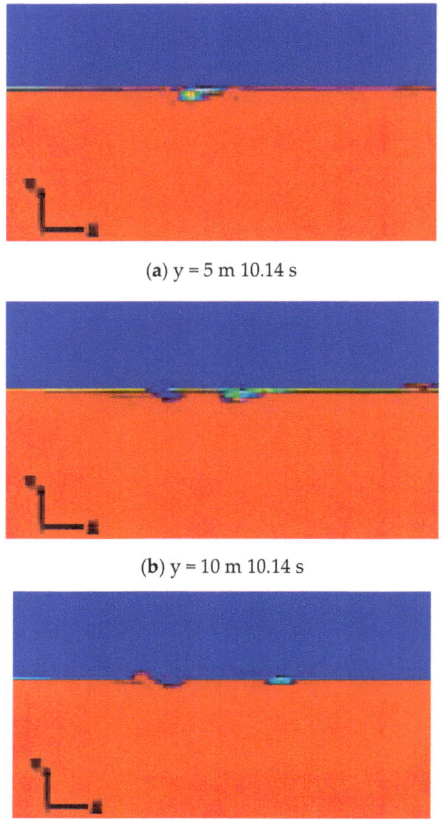

(a) y = 5 m 10.14 s

(b) y = 10 m 10.14 s

(c) y = 15 m 10.14 s

Figure 9. Three phase distribution of oil–water–air for icy waters at 10.14 s.

The width and thickness of the oil film are shown in Table 2. Note that the oil particles presented not only on the free surface but also below-the-water surface. This portion of oil gradually rose within the water until contacting the ice, and then spread to the surrounding water. The drift of the oil spill was evident in the simulation and the drift distance changed significantly. However, the change in the thickness of the oil film was not prominent. Therefore, it was concluded that the expansion mode of an oil spill in medium icy waters was dominated by drift motion.

Table 2. Numerical results of drift and diffusion of oil film.

Position of y/m	Time(t)/s	Width of the Oil Film/m	Thickness of the Oil Film/m
y = 5	2.34–10.14	2.67–6.46	1.82–1.99
y = 10	5.64–10.14	2.96–9.51	1.44–1.94
y = 15	9.48–10.14	2.72–4.10	1.23–1.34

3.5. *Interaction of Ice and Oil*

Considering current, wind and waves, the movement of ice floes with 6DOF included rolling, pitching and heaving motions on the water surface. The overset grid [31] technique was used to simulate the flow field of moving ice with large displacement. This technique simplified the construction of computational grids about complex geometries and allowed relative motion between embedded grids. Based on interpolating connections in every step in the time direction, at the state of ice floes flow associated with the non-linear wave, the background mesh established in the static domain was adopted to simulate the far field in water and the valid configurations of overset mesh were refined around boundaries consisting of ice particles.

In the oil-spill model, oil density and ice diameter were set at 840 kg/m^3 and 1.5 m, respectively. As shown in Figures 10 and 11, there were two main movements: (a) oil trapped under the ice floe while it rose to the surface and encountered the ice; (b) oil trapped by slush or brash (broken) ice and diffused to the surface of ice, which was consistent with the features of oil spreading in medium ice-density waters [9,11]. At the plane section of y = 0 m and y = 0.25 m, the oil drifted to 4.06 m at 3 s, and the horizontal and vertical displacement of the ice flow in the vicinity of the water surface were 1.61 m and 0.19 m, respectively. The pitch amplitude only increased from 0 to 6.0 deg. This is a small increase mainly affected by currents and waves. Due to the fact that the velocity in the *y* direction was limited to 10% of the wind velocity, the roll motion was completely negligible.

(**a**) y = 0 m 0.1 s

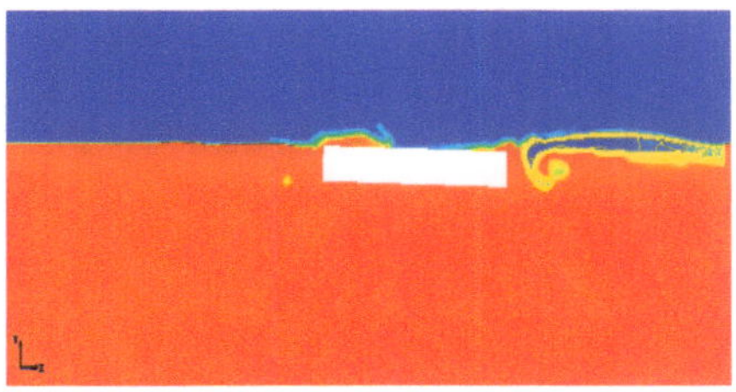

(**b**) y = 0 m 1 s

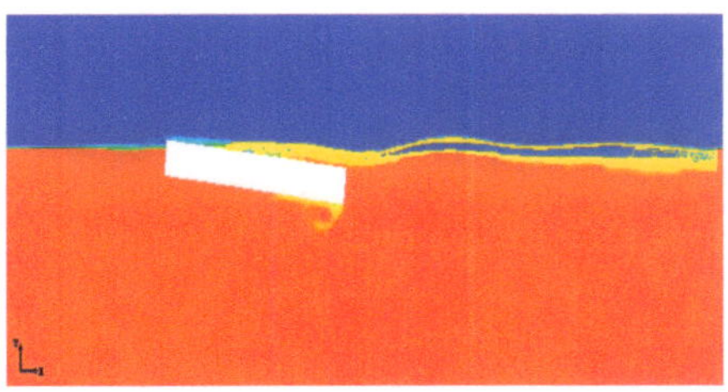

(**c**) y = 0 m 3 s

Figure 10. Oil-spreading characteristics of the ice particle with six-degrees-of-freedom (6DOF) movement at y = 0 m plane section.

(a) y = 0.25 m 0.1 s

(b) y = 0.25 m 1 s

(c) y = 0.25 m 3 s

Figure 11. Oil spreading characteristics of the ice particle with 6DOF movement at y = 0.25 m plane section.

4. Prediction of Oil-Spill Pollution Area

Once an oil-spill accident has occurred in icy waters, it is nearly impossible to clean up the oil by mechanical technology. Thereby, a low-temperature oil-dispersant could be adopted as the treatment method. The diffusion process of oil dispersants could be negligible because they attach themselves onto the surface of oil film [32]. However, the presence of fast-moving or drifting ice in Arctic waters will affect both oil and oil-dispersants irregularly during the spreading process. Therefore, it will be difficult to control the dosage of oil dispersants and might lead to severe damage to the Arctic Ocean environment. Accordingly, the changing extent of the oil-contaminated area in icy waters should be predicted as soon as possible in order to introduce effective emergency treatment.

For this purpose, approximate modeling technology utilizing a regression analysis to fit factors and responses was adopted. The RSM was used to establish an approximate model to depict the relationship between the responses for its benefits of simple mathematical expression, small calculation amount and fast convergence speed. The response surface in the form of polynomial function was used. The polynomial function response surface was verified with good fitting accuracy and calculation efficiency for the input–output relationship of most data structures. The relationship between factors and responses in the approximate modeling could be described by Equation (19).

$$\hat{y}(x) = a + \sum_{i=1}^{n'} b_i x_i + \sum_{i=1}^{n'} c_i x_i^2 + \sum_{i=1}^{n'-1} \sum_{j>i}^{n'} d_{ij} x_i x_j + \varepsilon \tag{19}$$

The polynomial function was adopted by RSM to fit the approximate model of the design space. The procedure for constructing the approximate model of the oil spill area used a third-order RSM, which was listed as follows:

1. The sample point x_1 was determined to construct the approximate model in the design space using the experimental design methodology, where $x_i = (x_1, x_2, x_3, \ldots, x_m)$, and m was the number of design variables. The data of sample points were the centroid coordinates (X, Y), area S, perimeter C', horizontal maximum distance L and vertical maximum distance D of the contaminated area, which were extracted from the numerical results of oil spill diffusion at different times as presented in Section 3.4 (i.e., 40 s, 80 s, 120 s, 160 s, and 200 s).
2. A series of sample pairs of designated variables and responses $\{(x_i, y_i)\}$ were computed by the model, where $y_i = (y_1, y_2, y_3, \ldots, y_n)$ was the response of the oil-spill area, and n was the number of sample points designed for the test.
3. The regression analysis using RSM was conducted to obtain an approximate model of the oil-spill area. The multiple correlation coefficient was adopted to evaluate the credibility of the response surface model. If the requirements were met, this model could be applied to the subsequent optimization computation. Otherwise, the credibility of the model could be improved by changing the model type and increasing the sample points.

The oil-spill area at a certain future time was predicted by the function that described the relationship between variables and responses. The appropriate equivalent weight of oil dispersants could be determined based on the oil-spill area to implement the emergency treatment. In order to shorten the calculation time, the prediction of the oil-spill pollution area was used as the continuation and supplement of the oil trajectory simulation. Accordingly, the response surface model could be established by using source data of the oil spill, which were attained from the numerical calculation results of Section 3 shown in Table 3. In order to improve the accuracy of fitting function, the functional relationship among t, (X, Y), C', L, and D was correlated by the fourth-order and sixth-order models. The design variable was t and the responses were X, Y, C', L, and D. The fourth-order and sixth-order models are shown in Equation (20).

$$\begin{aligned}
x &= 0.00000003t^4 + 0.0000006t^3 - 0.0027t^2 + 0.8342t + 94.618 \\
y &= -0.00000000005t^6 + 0.00000003t^5 - 0.000008t^4 + 0.0009t^3 - 0.0498t^2 \\
&\quad + 1.1752\mathrm{t} - 4.6404 \\
L &= 0.0000003\mathrm{t}^4 - 0.0001t^3 + 0.0077t^2 + 1.2593t + 97.143 \\
D &= -0.0000001\mathrm{t}^4 + 0.00002t^3 - 0.0291\mathrm{t} + 9.6553 \\
C' &= -0.000000002t^6 + 0.000001t^5 - 0.0003t^4 + 0.0243t^3 - 1.0821t^2 + 21.627\mathrm{t} \\
&\quad - 67.344
\end{aligned} \tag{20}$$

Table 3. Oil-spill parameters extracted from oil-spill trajectory simulation.

Time(t) /s	Area(S) /m^2	Perimeter(C′) /m	Centroid Centroid (X)/m	Coordinate Coordinate (Y)/m	Horizontal Maximum Distance (L)/m	Vertical Maximum Distance (D)/m
5	65.33	40.88	103.99	3.09	108.06	7.28
10	100.23	41.6	104.01	3.14	111.91	7.5
20	214.91	63.75	109.16	4.5	121.7	12.05
25	331.59	86.64	113.91	4.93	130.03	13.23
30	354.71	95.52	116.56	4.46	135.42	11.73
40	476.29	121.62	123	4.31	149.24	11.76
50	500.27	148.83	128.49	2.75	162.32	8.27
60	524.14	186.14	133.57	2.56	181.61	9.08
70	772.1	236.14	143.35	2.71	207.39	10.28
75	775.34	254.93	145.48	3.61	208.27	11.01
80	1138.79	302.18	154.09	4.98	224.39	15.93
90	1173.98	309.93	145.95	5.33	220.1	17.39
95	1262.44	346.96	147.97	6.16	220.89	21.58
100	1535.68	351.8	152.47	7.63	222.3	28.44
110	1618.11	408.45	151.36	7.93	222.72	30.97
120	2236.59	484.88	162.31	9.24	242.96	35.7
125	2291.43	454.33	168.08	8.13	251.08	32.12
130	2369.06	474.85	168.84	7.27	256.49	29.33
140	2423.48	432.39	176.93	6.63	266.22	24.62
150	2819.96	489.75	181.67	6.85	285.7	29.66
160	3344.54	690.65	173.01	11.14	270.35	55.5
170	3415.69	703.94	189.05	10.32	288	38.41
175	3674.69	883.42	190.71	11.51	306.49	45.38
180	3943.54	987.46	194.54	12.25	308.54	44.42
190	4225.53	942.32	202.43	10.35	338.16	41.04
195	4361.73	727.68	200.25	11.74	315.29	43.76
200	4792.95	707.69	207.89	10.7	340.31	41.7

The minimum number of sample points depended on the model order and the number of input variables. A third-order response surface model could be constructed for the oil-spill area S, for which the input variables were the perimeter C', centroid position (X, Y), horizontal maximum distance L and maximum vertical distance D. Thus, the number of variables M was 5. The minimum number of sample points must be equal to $(M+1)(M+2)/2+M$, which was 26 in this research. Therefore, the fitting equation of the oil-spill area S is shown in Equation (21).

$$\begin{aligned} S = 165789.84 \quad & +109.3C' - 4672.69X - 6503.4Y + 1138.7L + 973.86D + 0.18C'^2 + 36.87X^2 \\ & -349.34Y^2 + 1.26L^2 - 5.86D^2 + 0.1C' * X - 14.07C' * Y - 0.7C' * L \\ & +0.72C' * D + 72.93X * Y - 15.32X * L - 10.34X * D - 0.16Y * L + 37.99Y \\ & *D + 2.34L * D - 4.12C'^3 - 0.05X^3 + 26.73Y^3 + 0.01L^3 - 0.02D^3 \end{aligned} \tag{21}$$

The space model of factors and responses is shown in Figures 12 and 13. The local effects and global effects for responses were shown in Figures 14 and 15, respectively.

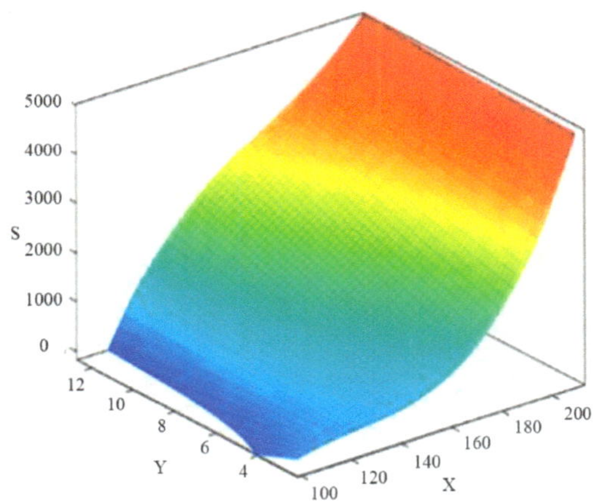

Figure 12. Three-dimensional space model of factors (X, Y) and responses (S).

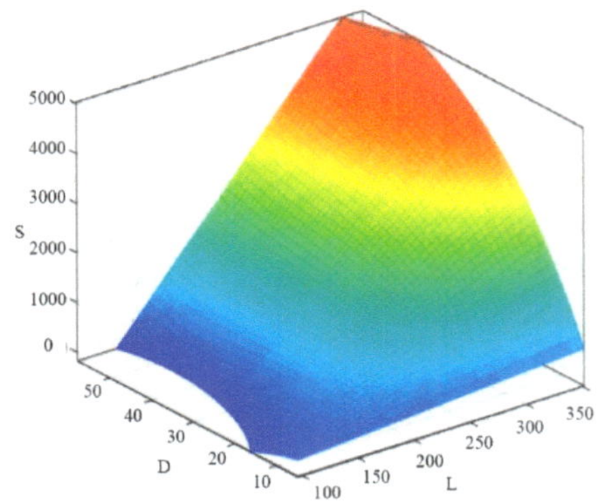

Figure 13. Three-dimensional space model of factors (L, D) and responses (S).

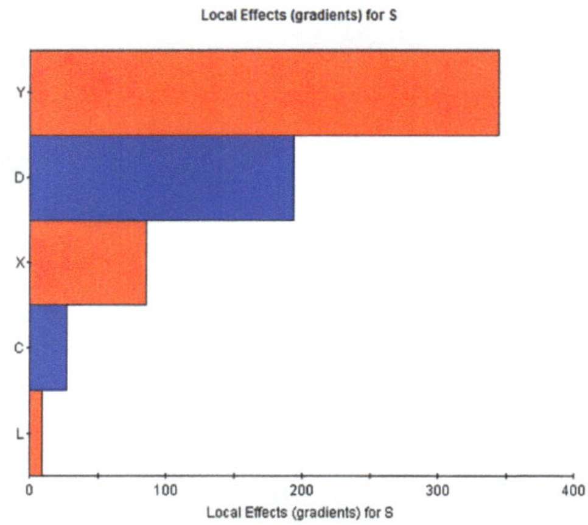

Figure 14. Local effects (gradients) for S.

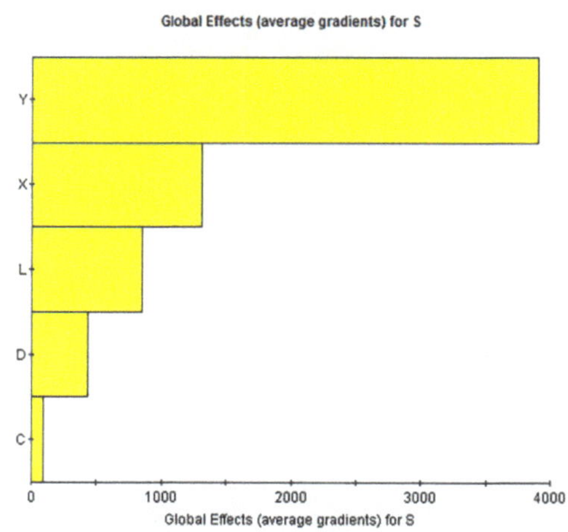

Figure 15. Global effects (average gradients) for S.

Obviously, factor Y played a dominant role in both local effects and global effects for S. Therefore, it was essential to predict the oil-spill area by designated centroid coordinate Y in icy waters. Emergency treatment measures for the oil spill could then be organized without considering non-sensitivity factors L and C'.

The reliability of this approximate model was evaluated by the multiple correlation coefficient R^2 as given in Equation (22):

$$R^2 = 1 - \sum_{i=1}^{n''} (y_i - \hat{y}_i)^2 / \sum_{i=1}^{n''} (y_i - \overline{y_i})^2 \tag{22}$$

The value of R^2 was used to measure the degree of conformity between the approximate model and the sample points. R^2 being close to 1.00 indicated that the approximate model had a relatively high degree of credibility. In this case, R^2 was 0.9 and achieved the established threshold, which meant the approximate model could provide a relatively credible prediction of oil-spill diffusion.

5. Conclusions

In this study, a three-dimensional DEM-based numerical model of ship oil-spill movement in Arctic icy waters was established. The trajectory model of oil film coupled with marine loading including wind, current and waves utilized the multiple tracking method of oil drift and ice floe drift. The accident scenario was simulated with reference to a real experiment. The interaction details of oil and ice floes were expanded by a 6DOF model to describe the features of oil spreading. Subsequently, the correlation of parameters on the behaviour of oil in cold water were examined by sensitivity analysis, and the area of the oil spill was predicted using an RSM model in order to conduct rapid emergency treatment. The RSM model was evaluated by reliability analysis to indicate a high degree of credibility.

This model was presumably based on the unique characteristics of the Arctic Ocean and combined effects of wind, wave, current and ice. The wind was regarded as a steady wind of winter; the wave was simulated as a fifth-order Stokes wave in non-linear motion over time; the current was the relatively stable polar snap coupled with the wave; the ice floe was sprayed for 100 s before the oil's spraying and stopped immediately when the oil started to spray. The oil spill was assumed as a continuous spill in a static point and the amount of oil spill was assumed to be 3 tons. The following conclusions were drawn:

1. The presence of ice floes had a significant suppression effect on the movement of the oil spill. The maximum drift distance of oil film and the maximum radius of oil diffusion in icy waters were found to be much smaller than those in open waters. With the combined action of wind, wave, current and ice, the oil film was observed to expand to a larger area in ice-free waters but form a long and narrow belt in icy waters.
2. The oil film diffusion and drift processes occurred simultaneously. The oil film in waters with medium-density ice was observed to be distributed on both the water surface and underwater. The underwater oil gradually rose until it came into contact with the ice, and then spread to the surroundings. The changes in the horizontal drift distance of the oil spill were more evident than those in oil-film thickness. Thus, oil-film drift constituted the primary behavior of oil-spill expansion.
3. The regression function of time t, the centroid coordinates (X, Y), perimeter C', horizontal maximum distance L and vertical maximum distance D were generated on the base of the approximate model method. This function exhibited a relatively high degree of credibility. It could predict the extent of oil spill in icy waters scientifically and reasonably, which could provide support for establishing a rapid emergency oil-spill cleanup system in icy waters, such as those found in the increasingly accessible Arctic Ocean.

Author Contributions: Formal analysis, W.L., X.L. and Z.D.; Methodology, W.L., X.L. and J.L. (Jianguo Lin); Supervision, J.L. (Jianguo Lin); Writing—original draft, W.L., P.G. and Q.M.; Writing—review and editing, J.L. (Jiamin Liu), H.W. and Z.S. All authors have read and agreed to the published version of the manuscript.

Funding: This research was funded by National Engineering Laboratory for Mobile Source Emission Control Technology (NELMS2018A06) and Fundamental Research Funds for the Central Universities of China (3132017097, 3132018177, 3132019304).

Conflicts of Interest: The authors declare no conflict of interest.

Abbreviations

DPM	Discrete Phase Model
DEM	Discrete Element Method
RSM	Response Surface Method
VOF	Volume of Fluid Method
6DOF	Six Degrees of Freedom

Notations

u	Velocity component of fluid in x direction
v	Velocity component of fluid in y direction
t	Flow time
F_q	Ratio of the volume of q-th phase fluid in the cell to the total cell volume
φ_q	Fluid parameter
F_{n}^{n}	Normal contact force at time n
K_{n}	Normal contact stiffness
C_{n}	Contact viscous coefficient
δ_{ij}	Overlap value between the two disks i and j
V_{ij}	Relative velocity vector of the contact faces
$\mathbf{n}$	Normal unit vector of the contact cells
F_{t}^{n}	Tangential forces at time n
F_{t}^{n-1}	Tangential forces at time $n-1$
K_{t}	Tangential stiffness between particles
Δt	Time step
$\mathbf{t}$	Tangential unit vector between contact cells
μ	Coefficient of sliding friction
C	Concentration
k	Vertical diffusion coefficient
K_H	Horizontal diffusion coefficient
P	Position of the marked substance micelle at time Δt
K	Normal coordinates
ΔA	Number of cells in the computational domain
V_i	Volume occupied by the particles on the ocean surface in grid cell
τ_x/f	Shear stresses of the wind in the x direction
τ_y/f	Shear stresses of the wind in the y direction
D_{s}	Diffusion coefficient of the oil film
f	Friction coefficient of the oil–water interface
g	Acceleration due to gravity
R_h	Physical–chemical kinetic terms
S_0	Initial position of the oil-film centroid
S	New position of the oil-film centroid after a time Δt
U_L	Velocity of oil-film drift
U_f	Velocity on water surface
U_w	Velocity of wind
a_f	Horizontal flow coefficient
a_w	Wind frag coefficient
x_0,y_0	Initial centroid position
x',y'	Centroid position after a time Δt
u	Current velocity in the x direction
v	Current velocity in the y direction
i_w	Correction factor
θ	Wind angle
$Mf\vec{k}\times\vec{U}$	Coriolis force
$A\vec{\tau_a}$	Force of the wind above ice floe
$A\vec{\tau_w}$	Force of the current above ice floe
$Mg\cdot gradH$	Effects of the ocean surface slope
$\vec{F}$	Non-linear ice internal force
H	Gravitational potential on the ocean surface
f	Coriolis force coefficient
$\vec{k}$	Unit vector perpendicular to the ocean surface
$\vec{U}$	Velocity of the ice floe
A	Ice density

h_a	Average ice thickness in the grid
M	Unit area mass of ice floe
Φ_h	Thermodynamic term determined by ice thickness
Φ_A	Thermodynamic term determined by ice density
$y(x)$	Response of a sample point
$\widetilde{y}(x)$	Response of an approximate model
x_i, x_j	Design parameter value
n'	Number of designated parameters;
ε	Error between the response value and true value
S	Area of oil film
C'	Perimeter of oil film
L	Horizontal maximum distance of oil film
D	Vertical maximum distance of oil film
n''	Sample numbers
y_i	Real value of simulation program
$\hat{y}_i$	Estimated value of response surface model
$\overline{y_i}$	Mean of true response

References

1. Pietri, D.; Soule, A.B.; Kershner, J.; Soles, P.; Sullivan, M. The arctic shipping and environmental management agreement: A regime for marine pollution. *Coast. Manag.* **2008**, *36*, 508–523. [CrossRef]
2. Eide, M.S.; Endresen, O.; Breivik, O.; Brude, O.W.; Ellingsen, I.H.; Røang, K.; Hauge, J.; Brett, P.O. Prevention of oil spill from shipping by modelling of dynamic risk. *Mar. Pollut. Bull.* **2007**, *54*, 1619–1633. [CrossRef]
3. Shiau, B.S.; Tsai, R.S. A numerical simulation of oil spill spreading on the coastal waters. *J. Chin. Inst. Eng.* **1994**, *17*, 473–484. [CrossRef]
4. Johansen, Ø. DeepBlow—A Lagrangian Plume Model for Deep Water Blowouts. *Spill Sci. Technol. Bull.* **2000**, *6*, 103–111. [CrossRef]
5. Yapa, P.D.; Li, Z. Simulation of oil spills from underwater accidents I: Model development. *J. Hydraul. Res.* **1998**, *35*, 673–688. [CrossRef]
6. Wang, P.; Li, Z.J. A Primary Numerical Model of Oil Spilling in Icing Sea. *J. Glaciol. Geocryol.* **2003**, *25*, 334–337.
7. Chen, X.; Yu, J.X.; Li, Z.G.; Wu, C.H.; Jiang, M.R.; Yang, Z.L. Application of Discrete Phase Model and Empirical Model in Oil Spill Simulation of Sea Surface. *China Offshore Platf.* **2019**, *34*, 47–53.
8. Ambjorn, C. Seatrack Web, Forecasts of Oil Spills, a New Version. *Environ. Res. Eng. Manag.* **2007**, *41*, 60–66.
9. Venkatesh, S.; El-Tahan, H.; Comfort, G.; Abdelnour, R. Modelling the behaviour of oil spills in ice-infested waters. *Atmos. Ocean* **1990**, *28*, 303–329. [CrossRef]
10. Wang, R.S.; Liu, Q.Z.; Chen, W.B.; Jin, H.T.; Liu, C.H. Numerical simulation and tests of sea ice drift process in the bohai sea. *Oceanol. Limnol. Sin.* **1994**, *3*, 10.
11. Xia, D.W.; Xu, J.Z. Investigation on numerical modelling of spill oil movement in sea ice infested waters. *Acta Oceanol. Sin.* **1998**, *1*, 113–122.
12. Huang, J.; Cao, Y.J.; Gao, S. Numerical simulation of oil spill drift-diffusion in the Bohai sea. *Mar. Sci.* **2014**, *38*, 100–107.
13. Huang, Y.; Song, M.R.; Guan, P. Study on the Varying Regularity of Oil Spill Properties During the Ice Period of the Bohai Sea. *J. Ocean Technol.* **2016**, *35*, 1–5.
14. Yu, J.A.; Zhang, B.; Liu, Q.Z.; Chen, W.B.; Wang, R.S. Numerical experiment on the behaviour of oil spills in ice-infested waters in the bohai sea. *Oceanol. Limnol. Sin.* **1999**, *30*, 557–562.
15. Wilkinson, J.P.; Wadhams, P.; Hughes, N.E. Modelling the spread of oil under fast sea ice using three-dimensional multibeam sonar data. *Geophys. Res. Lett.* **2007**, *34*, L22506. [CrossRef]
16. Xiao, M.; Gao, Q.J.; Lin, J.G.; Li, W.; Liang, X. Simulation of Submarine Pipeline Oil Spill Based on Wave Motion. In Proceedings of the International Conference on Computer Modeling and Simulation, Hainan, China, 22–24 January 2010; Volume 1, pp. 433–437.
17. Liu, X.F.; Meng, L.K.; Zhao, C.Y. DEM-based Oil and Gas Migration Pathway Simulation of Oil-and Gas-Bearing Basin. *Editor. Board Geomat. Inf. Sci. Wuhan Univ.* **2004**, *29*, 371–375.

18. Ma, E.; Xu, Z.M. Technological process and optimum design of organic materials vacuum pyrolysis and indium chlorinated separation from waste liquid crystal display panels. *J. Hazard. Mater.* **2013**, *263*, 610–617. [CrossRef]
19. Lv, G.L.; Liu, Z.M. Simulation and Analysis of Diffusion of Oil and Gas Leakage from Submarine Pipeline Based on CFD. *J. Jimei Univ.* **2017**, *22*, 51–57.
20. Ren, B.; Li, X.L.; Wang, Y.X. An Irregular Wave Maker of Active Absorption with VOF Method. *China Ocean Eng.* **2008**, *4*, 94–105.
21. Li, Z.G.; Jiang, M.R.; Yu, J.X. Numerical simulation on the oil spill for the submarine pipeline based on VOF method. *Ocean Eng.* **2016**, *34*, 100–110.
22. Li, Z.L.; Liu, Y.; Sun, S.S.; Lu, Y.L.; Ji, S.Y. Analysis of ship maneuvering performances and ice loads on ship hull with discrete element model in broken-ice fields. *Chin. J. Theor. Appl. Mech.* **2013**, *45*, 868–877.
23. Mohr, O. Welche Umstande bedingen die Elastizitatsgrenze undden bruch eines materials. *Z. Des. Ver. Dtsch. Ing.* **1900**, *44*, 1524–1530.
24. Egberongbe, F.O.; Nwilo, P.C.; Badejo, O.T. Oil Spill Disaster Monitoring along Nigerian Coastline. In Proceedings of the 5th FIG Regional Conference: Promoting Land Administration and Good Governance, Accra, Ghana, 8–11 March 2006.
25. Cheng, R.T.; Casulli, V.; Milford, S.N. Eulerian-Lagrangian Solution of the Convection-Dispersion Equation in Natural Coordinates. *Water Resour. Res.* **1984**, *20*, 944–952. [CrossRef]
26. Yang, Z.Y. B-Spline interpolation in Eular-Lagrangian method for convection-dispersion equation. *J. Ocean Univ. Qingdao* **1994**, *2*, 143–151.
27. Rasmussen, T.L.; Thomsen, E.; Kuijpers, A.; Stefan, W. Late warming and early cooling of the sea surface in the Nordic seas during MIS 5e (Eemian Interglacial). *Quat. Sci. Rev.* **2003**, *22*, 809–821. [CrossRef]
28. DF Dickins Associates. *Field Research Spills to Investigate the Physical and Chemical Fate of Oil in Pack Ice*; Environmental Studies Revolving Funds: Canada, 1987; 118p.
29. Sun, H.; Lu, P.; Li, Z.J. Numerical Simulation on Dynamic Characteristics of Flow Field under Ice. *Math. Pract. Theory* **2015**, *45*, 69–77.
30. Tu, X.C. Ice Resistance Prediction and Parameters Sensitivity Study for Polar Geophysical Prospecting Vessel. Master Thesis, Jiangsu University of Science and Technology, Zhenjiang, China, 2019.
31. Ma, S.; Ge, W.P.; Duan, W.Y.; Liu, H.X. Simulation of free decay roll for C11 container ship based on overset gird. *J. Huazhong Univ. Sci. Technol.* **2017**, *45*, 34–39.
32. Li, Y.; Liu, B.X.; Lan, G.X.; Ma, L.; Gao, C. Study on Spectrum of Oil Film in Ice-Infested Waters. *Guang Pu Xue Yu Guang Pu Fen Xi* **2010**, *30*, 1018–1021.

© 2020 by the authors. Licensee MDPI, Basel, Switzerland. This article is an open access article distributed under the terms and conditions of the Creative Commons Attribution (CC BY) license (http://creativecommons.org/licenses/by/4.0/).

applied sciences

MDPI

Article

A Numerical Ice Load Prediction Model Based on Ice-Hull Collision Mechanism

Meng Zhang [1], Karl Garme [1], Magnus Burman [1] and Li Zhou [2,*]

1 Department of Vehicle Engineering and Solid Mechanics, KTH Royal Institute of Technology, Box 24075, 10450 Stockholm, Sweden; mengzh@kth.se (M.Z.); garme@kth.se (K.G.); mburman@kth.se (M.B.)
2 Department of Naval Architecture and Ocean Engineering, Jiangsu University of Science and Technology, Zhenjiang 212003, China
* Correspondence: zhouli209@hotmail.com; Tel.: +86-152-6291-0667

Received: 18 December 2019; Accepted: 14 January 2020; Published: 19 January 2020

check for updates

Abstract: A simplified numerical model is introduced to predict ice impact force acting on the ship hull in level ice condition. The model is based on ice-hull collision mechanisms and the essential ice breaking characteristics. The two critical ice failure modes, localized crushing and bending breaking, are addressed. An energy method is used to estimate the crushing force and the indentation displacement for different geometry schemes of ice-ship interaction. Ice bending breaking scenario is taken as a semi-infinite plate under a distributed load resting on an elastic foundation. An integrated complete ice-hull impact event is introduced with ice failure modes and breaking patterns. Impact location randomness and number of broken ice wedges are considered in order to establish a stochastic model. The analysis is validated by comparison with the model ice test of a shuttle passenger ferry performed in May 2017 for SSPA Sweden AB at Aker Arctic Technology Inc. Good agreement is achieved with appropriate parameter selection assumed from the model test and when ice bending failure is dominant. This model can be used to predict the ice impact load and creates a bridge between design parameters (ice properties and ship geometry) and structure loads.

Keywords: ice-hull interaction; level ice; ice load predication; model validation; ice model test; parametric study

1. Introduction

According to the European Commission, inland waterway transportation (IWT) is classified as one of the five transport modes. It can release traffic congestion and reduce greenhouse gas emissions. For countries with strong winters, the development of IWT must take into account the impact of ice problems on ice-going ships. The most common problems are the ship resistance in icy waters and impact ice loading on the hull. There are many available empirical methods [1–3] for ice-induced resistance calculation. Hence the problem could as an initial step be solved by increasing the propulsion power. For ice impact load, several methods can be used. In principle, ice-hull collision occurs in the bow area. Therefore, it is critical to put focus on the bow structure in the design process and to study the hull design parameters and their influences on the ice impact load.

One way is to utilize the probabilistic method which simplifies the ice pressure in relation to the contact area,

$$\alpha = Ca^{D}, \tag{1}$$

where C and D are constants for a given ice condition, α is pressure on structure in [MPa] and a is contact area between ice floe and ship structure in [m^2]. The probabilistic method aims to estimate extreme ice load and is easy to apply compared to other methods. However, this method is proposed for sea-ice going ships with severe ice conditions and the existing values of the two constants come from sea trial tests [4,5]. The inland waterway (IWW) ice condition is however drastically different compared to the sea ice condition and the discrepancies in ice properties are significant, resulting in a different scale of ice loading. It can result in extreme high pressure for light ice condition. The ship design with regards to the corresponding ice predication is too conservative and not efficient. Moreover, concerning IWW ice conditions, there are uncertainties due to the lack of research and test data and that make it difficult to estimate the ice properties and ice loads accurately [6]. Nevertheless, it is still a good approach to estimate an extreme load even when several design parameters are unknown. Practically, when it comes to the ship design, more parameters and their effects on ice load are usually required. However, the direct ice load prediction needs more refined models as there are lots of influencing parameters both from hull geometry and ice properties.

Another approach is to consider the physical phenomena and reduce the complex scenario to single events by studying the ice-hull interaction mechanism and ice failure modes (e.g., crushing, bending, splitting, sliding). The studies based on ice-hull interaction process are more reliable and efficient for practical ice collision scenarios and ship design aspects. Studies, presented in the literature available either study ice crushing or ice bending, but the combination of the two failure modes are limited. Investigating the two modes together has a significant meaning. By considering the two failure modes simultaneously, a more reliable ice load prediction model should be obtained.

In this work, hull-ice impact is studied, and the focus is on both of the ice crushing and ice bending case, with the contact area connecting the two modes. Normally, ice bending failure is widely accepted as the dominant ice breaking pattern with the largest force. The same assumption is applied to the current model to ensure the accuracy of the model.

The purpose of the paper is to propose a feasible and reliable model for ice load prediction, especially the ice impact force, F_i on ship hull. The energy method [7] is chosen as to calculate the crushing force and the contact area. The stem collision and the shoulder collision are evaluated with two different contact area cases respectively. The bending scenario is equivalent to a wedge-shaped beam failure model and is represented by a semi-plate resting on a Winkler foundation with the ice flexural strength, σ_f as the failure criterion. There are also other empirical equations widely used for predicting the ice bending breaking force, for example, $P_b = C{\cdot}\sigma_f{\cdot}{h_i}^2$ in which h_i is the ice thickness and C is a constant based on previous research [8]. However, only ice flexural strength and ice thickness are taken into consideration. The bending scenario used in this paper is more suitable since many design parameters are considered. A parametric study is performed with the aim to reveal influences from the design parameters. The numerical model is constituted by repetition of a complete collision event with a set of ice failures happening in sequence.

2. Ice Load Prediction Model

The ship-ice interaction process is known as a complex phenomenon which highly depends on ice properties, hull geometry and relative speed [9]. Valanto [10] divided this interaction process into ice breaking, ice rotation, ice sliding, and ice clearing based on the main phenomena in breaking the level ice [11]. This break-displace process and ice forces are idealized in Figure 1.

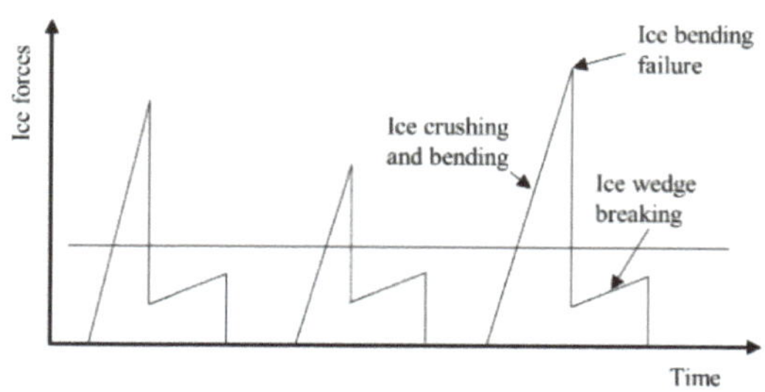

Figure 1. Idealized time histories of ice forces.

The ship-ice interaction begins with a localized crushing of a free ice edge at the contact zone. When the ship keeps moving forward, both the contact area and the crushing force increase. The ship advance motion causes the ice sheet to deflect down along the hull surface, thus bending stresses are build up in ice sheet until it breaks off into small pieces, as shown in Figure 2. The ice flexural failure occurs at a distance, expressed as the ice breaking length, l_b from the crushing region. It is mainly determined by the ice thickness and the ship speed among all other parameters. The final force to break the ice sheet is denoted as the ice breaking force P_b.

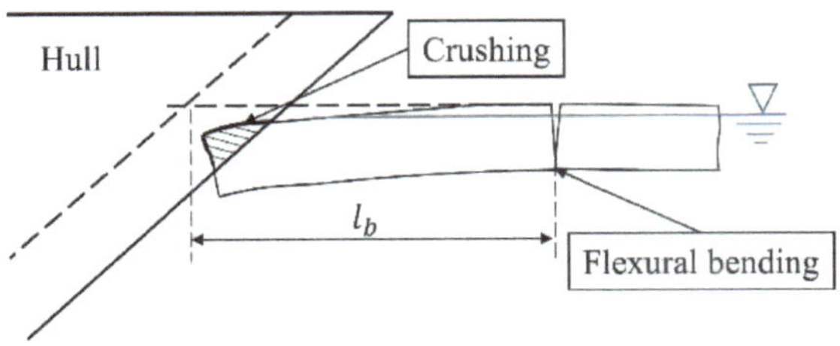

Figure 2. Illustration of the ice breaking model between a ship bow and level ice.

The numerical model developed in this paper aims to provide an efficient way to compute ice-hull impact force with two primarily ice failure features. Ice is regarded as an isotropic material with flexural strength, σ_f which is used as the failure criterion. The process is assumed to be a repetition of the following steps: (1) initial contact and crushing, (2) cracking due to ice bending, (3) more cracks are triggered and both radial and circumferential cracks are generated. A detailed process of the periodic ice impact load prediction model and calculation procedure are well described herein.

2.1. *Ice-Hull Contact Mechanism and Determination of Impact Load*

Four ice failure occurrences are assumed to happen during one complete impact event. The forces for each occurrence can be computed. The time history ice force is thus a periodic function of every impact event.

2.1.1. Ice Crushing

Ice crushing is regarded to occur at the first stage during one impact event. Ship penetrates ice sheets with a constant speed. An ice sheet is unbreakable but localized fracture takes place with ice spalling and extruding. The crushing force, F_C, indentation displacement, ζ_n, and contact area, A_n, are therefore determined by the energy method.

2.1.2. Initial Crack Propagation

The initial ice sheet defect due to bending failure usually happens along the free edge of ice sheet and the radial fracture is roughly perpendicular to the free boundary. A similar phenomenon is considered by Lu et al. [12] and is denoted as ice splitting failure. This occurs if ice floes are present with finite mass. The current case is applicable for infinite ice sheet. Shear force is not considered in ice bending as it is proven trivial. Thus, ice breaks when $\sigma_{yy}(0,0) \geq \sigma_f$ on the ice sheet free edge. An assumption is made that initial crack happens before the maximum stress is build up and corresponding force is not considered in the model.

2.1.3. Circumferential Ice Breaking

Circumferential cracks are assumed to form after any initial defect is created. The ice sheet starts to break up due to bending failure occurring at the maximum stress, $\sigma_{\max}$. In the ice bending breaking model, with the failure criterion, $\sigma_{xx\_max} \geq \sigma_f$, the position of the maximum stress can be located and the maximum ice sheet breaking force, P_{bmax}, can be found. The ice breaking length, l_b, as shown in

Figure 2, occurs at a point where the maximum stress just exceeds the ice flexural strength. Here the computed value of ice breaking length from ice bending breaking model is denoted as l_{b1},

$$l_{b1} = x \; at \; \sigma_{\max} = \sigma_f, \tag{2}$$

There are other empirical ways to calculate the ice breaking length. Usually it is proportional to the ice characteristic length, l_c. Zhang et al. [6] assumed that the breaking length l_b is one-third of l_c. This assumption originates from the ice resistance prediction method proposed by Lindqvist [1]. The empirical ice breaking length value is denoted as l_{b2}.

$$l_{b2} = \frac{1}{3} l_c, \tag{3}$$

$$l_c = \left(\frac{E {h_i}^3}{12(1 - \nu^2) \rho_w g} \right), \tag{4}$$

where ρ_w is the water density, E is the ice Young's modulus, h_i is ice thickness and ν is the ice Poisson's ratio.

2.1.4. Wedge Shaped Ice Floe with Radial Cracks

Eventually more radial cracks are produced as the floating ice wedges are broken off from the ice sheet. As introduced by Kerr [13], the failure load for a wedged ice with an opening angle θ can be calculated according to:

$$P_b = C_f \left(\frac{\theta}{\pi} \right)^2 \sigma_f h_i^2, \tag{5}$$

where C_f is an empirical parameter and donates the magnitude of ice breaking force.

Kerr [13] related the ice failure load to the maximum ice force by multiplying a square of the ratio of opening angle θ to π. Here the same analogy is applied. An expression is introduced to correlate the failure load of a wedged ice floe with a maximum ice breaking force, which is written as:

$$P_b = P_{bmax} \left(\frac{\theta}{\pi} \right)^2. \tag{6}$$

2.2. Stochasticity and Discretization in the Present Model

Ice-hull impact is a highly stochastic process. Normally, ice collision happens along the ship waterline plane randomly. Still, most impacts occur at the ship bow part which means that the stem and shoulder are more effective in terms of ice breaking. To estimate the ice breaking force, the periphery of waterline plane is discretized into small elements and the ice breaking force is computed in the time domain. In order to simplify the ice-hull interaction model and keep the dynamic feature, the ship stem and bow parts are highlighted in the model and discretized with infinite number of elements (points). Thus, several contact locations within the target area are evaluated in the model, as shown in Figure 3. In this paper, only five points are selected to demonstrate the concept. More points (elements) can be introduced to increase the model accuracy. In the model, the number of collisions is also randomized from one location to four locations for shoulder collision event.

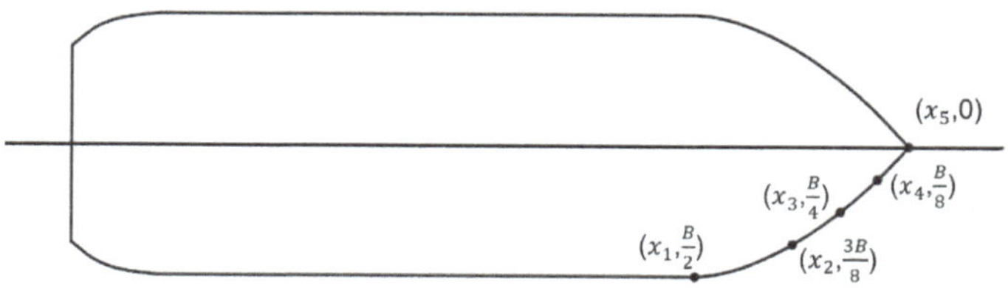

Figure 3. Bow discretization and locations considered in the paper.

When ice sheet breaks, it may break into different number of pieces. In this paper, the ice sheet is assumed to break into small ice floes with the same angle and size. Here the number of ice floes evaluated are 1, 2, 3 and they correspond to wedge angle of 180°, 90°, 60°. Those parameters are randomly selected in the simulation. A random wedge angle, e.g., 35°, can be explored in further study. All the scenarios mentioned above are considered to happen randomly in the model.

2.3. Model Implementation

The model presented is this paper consists of ice crushing force and bending force at different scenarios. Extra attention is paid to when the maximum ice sheet breaking force is computed. The process is illustrated in Figure 4. The whole computation is realized by programming in Matlab.

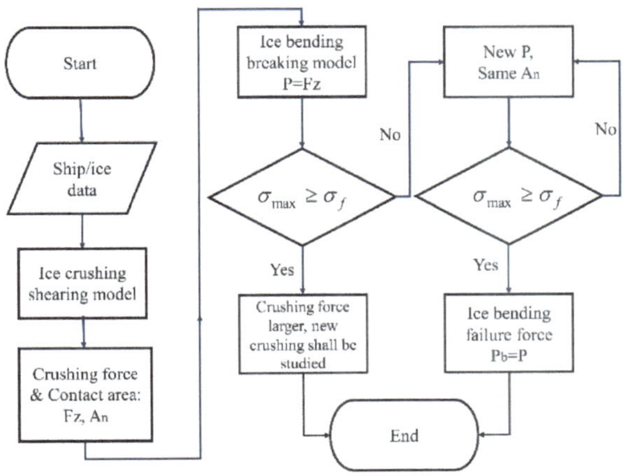

Figure 4. Description of how to obtain ice breaking force in the current model.

Traditional numerical methods require computation at each time step. Consequently, the simulation usually takes substantial computation resource and time. Instead of running heavy computation, the impact forces are here evaluated at different locations and scenarios, after that the force history can be generated by introducing the randomness and following the collision mechanism. Computation steps are used rather than the real event time since the final force is not a function of time. Although it is possible to refine the model using the real time, the computation then becomes more tedious.

3. Ice Load Prediction Sub Models

The fundamentals of the paper are based on the understanding and features of the ship-ice interaction described above. Currently, models evolved on such interaction are widely used in predicting the ice resistance, Lindqvist [1], Keinonen et al. [2], Riska et al. [3], and Su et al. [14]. In addition, this interaction decomposition offers a useful tool to predict the structural loads as well. Therefore, the model in this paper explores the potential to determine the structure load as well. This method would assist engineers to predict the ice load in an easy way and facilitate the hull structure design in accordance with ice loads and design parameters.

3.1. Ice Crushing-Shearing Model

The energy method proposed by Daley [7] and Propov et al. [15] provides an effective way to calculate the crushing force and indentation displacement for any ice-hull contact. The initial impact collision is triggered by a ship with kinetic energy and a stationary ice sheet with bulky mass. The inputs for calculation are the ice properties, e.g., ice thickness, ice crushing strength, and the hull geometry. Moreover, the method includes analytical formulas for the different geometric contact cases.

The hull geometry parameters are described and presented in Table 1 and Figure 5. The definitions of the hull angles from literature and Finnish-Swedish Ice Class Rules (FSICR) [16] are summarized and illustrated in Figure 6. The coordinate system is set with the origin at the mass center of a ship. The x-axis is along the longitudinal direction of the hull and positive forward. The y-axis is normal to the x-axis and is positive towards starboard. The z-axis is perpendicular to the xy-plane and positive downward [17].

Table 1. Ship parameters.

Parameters	Description	Parameters	Description
L	Waterline plane length	β	Frame angle
B	Waterline plane beam	B'	Normal frame angle
T	Draft	γ	Sheer(buttock) angle
D	Depth	φ^{*} [1]	Rake angle
x, y, z	The collision point on the hull or the ice sheet	C_b	Block coefficient
cg	Centre of gravity	C_{wp}	Water plane area coefficient
α	Waterline angle	C_m	Midship section coefficient

[1] when the buttock is measured on the ship hull away from x-axis with a distance of B/4, φ^* is denoted as φ_2; similarly, φ_1 is the ship stem angle when measured at ship central line.

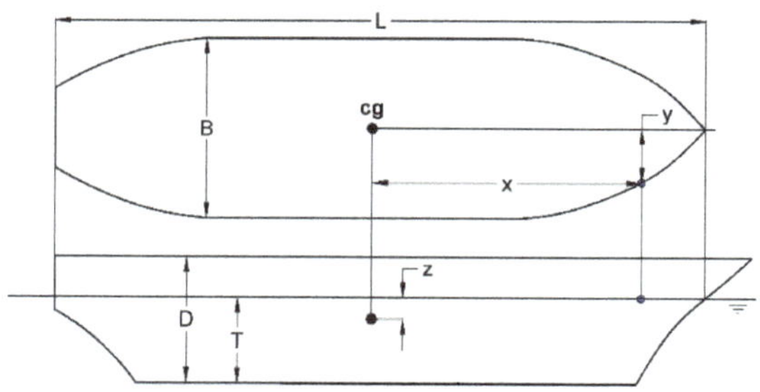

Figure 5. Definitions of geometric parameters at a ship-ice collision point.

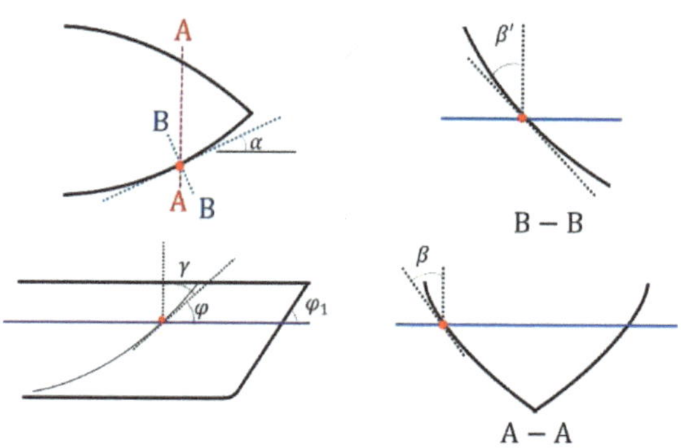

Figure 6. Definitions of hull angles at a collision point.

The relationships between ship angles are given as below:

$$\tan\beta = \tan\alpha\tan\gamma, \tag{7}$$

$$\tan\beta' = \tan\beta\cos\alpha. \tag{8}$$

The energy method is applicable for an impact between two objects. It is assumed that one object is moving while another is still. During the process, the effective kinetic energy from the moving object is equivalent to the energy expanded to the another as crushing, or indentation, energy. The collided object is assumed to have infinite large mass, thus the effective kinetic energy is regarded the same as the total kinetic energy. This approach is used to determine the maximum force when all motions are ceased.

According to such assumption and based on Daley [7], the effective kinetic energy, KE_e, and the crushing energy, IE, are equal:

$$KE_e = IE. \tag{9}$$

The indentation energy can be represented by the indentation force, F_n, and the crushing indentation displacement, ζ_c; the kinetic energy is expressed by the ship effective mass and velocity:

$$IE = \int_0^{\zeta} F_n d\zeta_c, \tag{10}$$

$$KE_e = \frac{1}{2} M_e {V_n}^2, \tag{11}$$

$$M_e = \frac{M_{ship}}{C_o}, \tag{12}$$

$$V_n = V_{ship} l, \tag{13}$$

where M_e is the effective mass at the impact point; M_{ship} stands for the effective mass of the ship at the contact point; C_o presents the mass reduction factor in six degrees of freedom; V_n is the normal velocity at the impact point; V_{ship} is the ship velocity; l is the x-direction cosine. They are defined based on Popov [15]:

$$l = \sin(\alpha)\cos(\beta'),\ m = \cos(\alpha)\cos(\beta'),\ n = \sin(\beta'). \tag{14}$$

For a symmetrical collision on the stem, the relationships become:

$$l = cos(\gamma),\ m = 0,\ n = sin(\gamma). \tag{15}$$

The roll, pitch, and yaw lever arm can be calculated, respectively:

$$\lambda_1 = ny - mz,\ \mu_1 = lz - nx,\ \eta_1 = mx - ly. \tag{16}$$

The added mass terms in surge, sway, heave, roll, pitch, and yaw can be obtained from [15]:

$$\begin{aligned} AMx &= 0, \\ AMy &= \tfrac{2T}{B}, \\ AMz &= \frac{2BC_{wp}^2}{3T\left(C_b\left(1+C_{wp}\right)\right)}, \\ AMrol &= 0.25, \\ AMpit &= \frac{B}{T\left(3-2C_{wp}\right)\left(3-C_{wp}\right)}, \\ AMyaw &= 0.3 + \tfrac{0.05L}{B}. \end{aligned} \tag{17}$$

The squared mass radii of gyration in roll, pitch and yaw can be expressed as:

$$\begin{aligned} rx^2 &= \tfrac{C_{wp}B^2}{11.4C_m} + \tfrac{D^2}{12}, \\ ry^2 &= 0.07C_{wp}L^2, \\ rz^2 &= \tfrac{L^2}{16}. \end{aligned} \tag{18}$$

The mass reduction coefficient C_o thus can be obtained [15]:

$$C_o = \frac{l^2}{1+AMx} + \frac{m^2}{1+AMy} + \frac{n^2}{1+AMz} + \frac{\lambda_1^2}{rx^2(1+AMrol)} + \frac{\mu_1^2}{ry^2(1+AMpit)} + \frac{\eta_1^2}{rz^2(1+AMyaw)}. \tag{19}$$

The normal impact force F_n and the energy IE can be calculated as:

$$F_n = p_o \cdot fa \cdot \zeta_n^{fx-1}, \tag{20}$$

$$IE = p_o \cdot fa \cdot \frac{\zeta_n^{fx}}{fx}, \tag{21}$$

where p_o is the pressure on 1 m^2, and equivalent to the ice crushing strength, σ_c, it equals to 2000 kPa ζ_n is the ice indentation from the initial contact point along the normal direction of contact area and fx is a function of ex, where ex is constant. fa is a function of the geometric parameter.

Based on Equations (8) and (20), the kinetic energy can be expressed as:

$$KE_e = p_o \cdot fa \cdot \frac{\zeta_n^{fx}}{fx}, \tag{22}$$

with the normal indentation displacement ζ_n as:

$$\zeta_n = \left(\frac{KE_e \cdot fx}{p_O \cdot fa} \right)^{\frac{fx-1}{fx}}, \tag{23}$$

Substitute Equation (22) into Equation (19), the normal force can be calculated:

$$F_n = p_O \cdot fa \cdot \left(\frac{KE_e \cdot fx}{p_O \cdot fa} \right)^{\frac{fx-1}{fx}}. \tag{24}$$

It can be observed that the indentation displacement ζ_n plays an important role. It is affected by the impact geometry configuration. Two indentation geometry scenarios are introduced. They refer to the bow stem (or head on) collision and the bow shoulder collision respectively, as seen in Figures 7 and 8. The two collision scenes are illustrated as a symmetrical- and an unsymmetrical-contact, as shown in Figures 7 and 8. The indentation functions, fx and fa are defined in Daley et al. [15]. The ice sheet dimension and bow geometry at the collision area shown in Figure 9, and the indentation areas for the two collisions and two contact area cases can be derived based on the indentation functions [15].

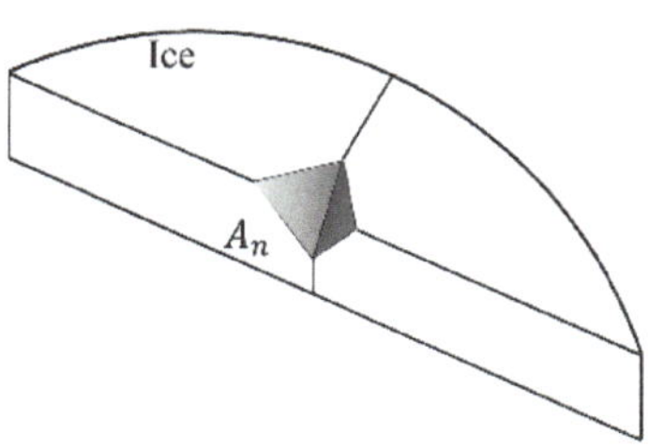

(a)

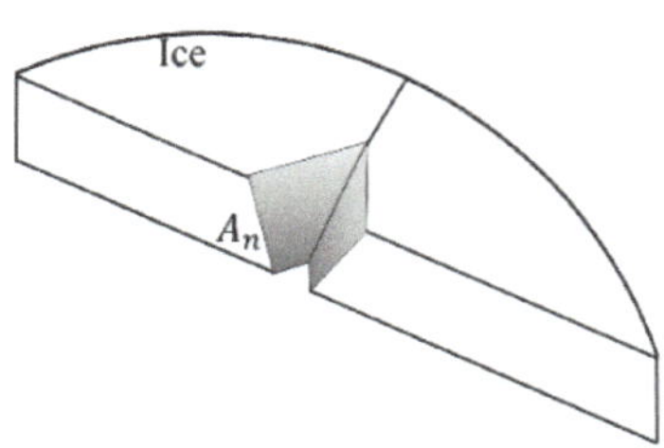

(b)

Figure 7. Case I-bow front collision: symmetric V wedge indentation. (**a**) Indentation does not go through the ice thickness; (**b**) Indentation penetrates through the ice thickness.

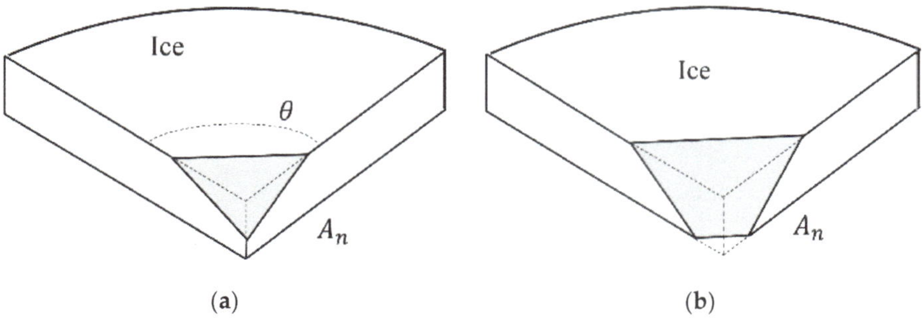

Figure 8. Case II-shoulder collision: wedge-shaped edge indentation. (**a**) Indentation does not go through the ice thickness; (**b**) Indentation penetrates through the ice thickness.

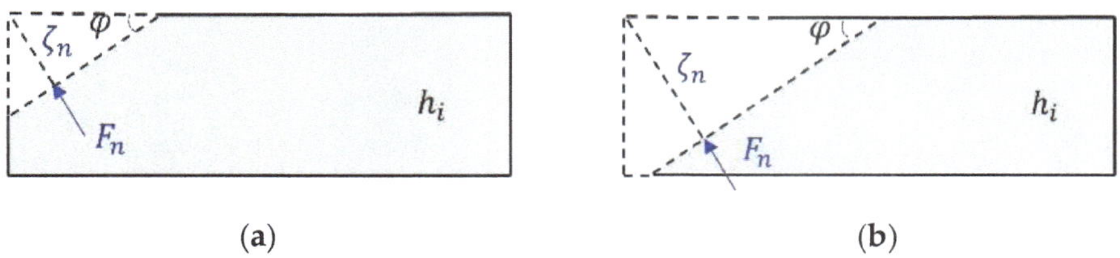

Figure 9. Side view of the two contact areas: (**a**) Case I with an indentation that does not go through the ice thickness *hi*; (**b**) Case II with an indentation penetrating through the ice thickness *hi*.

Collision I: bow front collision,

$$A_n = \begin{cases} \zeta_n^2 \frac{\tan\alpha}{\sin^2\varphi^* \cos\varphi^*}, & 0 \le \frac{\zeta_n}{\cos\varphi^*} \le h_i \\ \frac{2\zeta_n - h_i{}^2 \cos\varphi^* \tan\alpha}{\sin^2\varphi^*}, & \frac{\zeta_n}{\cos\varphi^*} \ge h_i \end{cases} \tag{25}$$

Collision II: shoulder collision,

$$A_n = \begin{cases} \frac{\zeta_n^2}{\cos^2\beta\prime \sin\beta\prime}, & 0 \le \frac{\zeta_n}{\sin\beta\prime} \le h_i \\ \frac{2\zeta_n h_i - h_i{}^2 \sin\beta\prime}{\cos^2\beta\prime}, & \frac{\zeta_n}{\sin\beta\prime} \ge h_i \end{cases} \tag{26}$$

The normal contact force can be divided into two components, the vertical component, *Fz*, and the horizontal component, *Fx* [18]. In the presented model, these force components are related to the failure criterion of the ice floe and the ice resistance, respectively. The vertical component, *Fz*, could cause bending failure of the ice floe. *Fz* can be calculated according to the relationship in Figure 10, and based on the geometric relationships, the resistance component and bending force can be calculated and transformed.

$$F_z = F_n(\cos\varphi - \mu\sin\varphi), \tag{27}$$

$$F_x = F_n(\sin\varphi + \mu\cos\varphi), \tag{28}$$

where *Fx* is the horizontal force, μ is the average dynamic coefficient of friction between the ice and the hull.

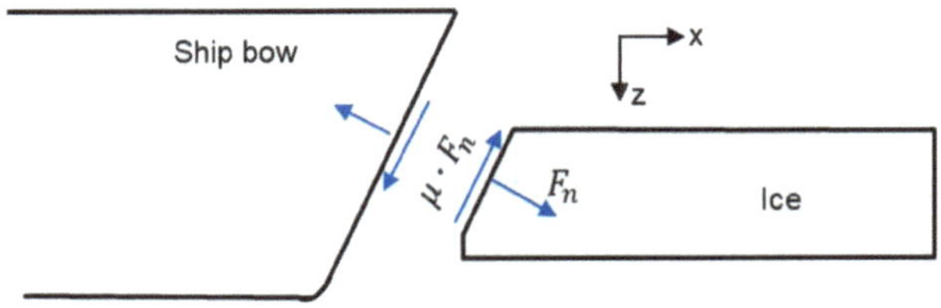

Figure 10. The force components at contact area.

3.2. Ice Bending-Breaking Model

The problem of computing the stresses in a breakable ice floe caused by ice-hull impacts can be solved by reforming it into an infinite plate subjected to a uniform and distributed force on a Winkler (elastic) foundation, as shown in Figure 11. The stress in the ice wedges can be calculated using method described in the literature [19–22].

The analysis is based on the differential equation:

$$D\nabla^4\omega + k\omega = q. \tag{29}$$

where D is the bending stiffness of the ice, ω is the vertical deflection, k is the specific weight of water (is the foundation modulus for a pavement or the specific weight of the liquid base), and q is the distributed load, with

$$\begin{aligned} D &= \frac{0.42Eh_i^3}{12(1-v^2)}, \\ k &= \rho_w g, \\ \nabla^4 &= \frac{\partial^4}{\partial x^4} + 2\frac{\partial^4}{\partial x^2 \partial y^2} + \frac{\partial^4}{\partial y^4}, \end{aligned} \tag{30}$$

where E is Young's modulus in [Pa], v is Poison's ratio, ρ_w is water density in $\left[\text{kg/m}^3\right]$, g gravitational constant in $[\text{N} \cdot \text{m}^2/\text{kg}^2]$.

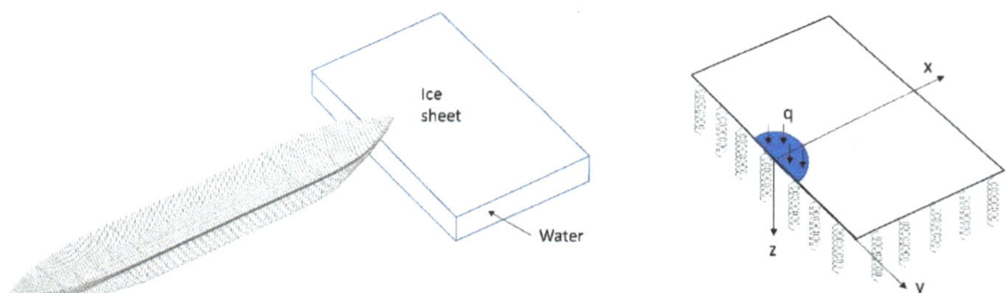

Figure 11. The ice bending model illustration: a semi-infinite plate resting on an elastic foundation and subjected to a uniform distributed load.

It can be seen from Figure 11 that the applied force is uniformly distributed over a half circular area with a radius r. As it is assumed that the contact area remains the same after ice crushing, the loading radius and distributed load can be computed as:

$$r = \sqrt{\frac{2A_0}{\pi}}, \tag{31}$$

$$q = \frac{F_z}{A_h} \text{ or } q = \frac{P}{A_h}, \tag{32}$$

where A_h is the horizonal area in the crushing zone. When the crushing force F_z is not sufficient to break the ice sheet, a new variable P is introduced to find the ice bending breaking force. The final breaking force can be found by iteration and is represented as P_b.

To solve the special case in Equation (29), the problem can be further reformulated from a uniform distributed load to a concentrated load acting close to the ice edge. Based on the work by Lubbad and Løset [9], the point load location can be marked as $P\ (x_0, 0)$, where

$$x_0 = \frac{4r}{3\pi}. \tag{33}$$

The general solution of the differential equation is as follows;

$$w = X_1 + X_2, \tag{34}$$

$$X_1 = \frac{-P\beta^2}{2\pi k}[\ker(\beta r_1) + \ker(\beta r_2)], \tag{35}$$

$$X_2 = \int_0^{\infty} A_1 e^{-\rho x}[\cos(x) - \rho^* \sin(x)] \cos(\alpha y) \mathrm{d}x, \tag{36}$$

where

$$\begin{aligned} A_1 &= \frac{P\beta^2}{\pi k} \cdot \frac{\mathrm{e}^{-\varrho x_0}}{\sqrt{\alpha^4+\beta^4}} \cdot \frac{2\varrho^2-(1-\nu)\alpha^2}{4\varrho^2[^2+(1-\nu)\alpha^2]-(1-\nu)^2\alpha^4} \cdot \left[\left(^2 + \varrho^2 + \nu\alpha^2\right) \cdot \cos(x_0) - \right. \\ &\quad \left. \varrho\left(^2 + \varrho^2 - \nu\alpha^2\right) \sin(x_0)\right], \\ r_1 &= \sqrt{(x - x_0)^2 + y^2}, \\ r_2 &= \sqrt{(x + x_0)^2 + y^2}, \\ \beta &= \sqrt[4]{k/D}, \\ \varrho &= \sqrt{\tfrac{1}{2}\sqrt{\alpha^4 + \beta^4} + \tfrac{1}{2}\alpha^2}, \\ &= \sqrt{\tfrac{1}{2}\sqrt{\alpha^4 + \beta^4} - \tfrac{1}{2}\alpha^2}, \\ \varrho^* &= \frac{\varrho[2^2+(1-\nu)\alpha^2]}{[2\varrho^2-(1-\nu)\alpha^2]}. \end{aligned} \tag{37}$$

The bending moments and stresses in the x- and y-direction are calculated from the curvature of the deformed ice sheet along x-axis at $y = 0$.

$$M_{xx}(x,0) = -D\left(\frac{\mathrm{d}^2 w}{\mathrm{d}x^2} + \nu\frac{\mathrm{d}^2 w}{\mathrm{d}y^2}\right), \tag{38}$$

$$M_{yy}(x,0) = -D\left(\nu\frac{\mathrm{d}^2 w}{\mathrm{d}x^2} + \frac{\mathrm{d}^2 w}{\mathrm{d}y^2}\right), \tag{39}$$

$$\sigma_{xx}(x,0) = -6\frac{M_{xx}}{h_i^2}, \tag{40}$$

$$\sigma_{yy}(x,0) = -6\frac{M_{yy}}{h_i^2}. \tag{41}$$

Moreover, the stress at the breaking point along the ice sheet free edge can be obtained by Lubbad and Løset [9] as

$$\sigma_{yy}(0,0) = 3qrD(1+\nu) \cdot kei(\beta r) + \frac{6D}{h^2} \cdot \int_0^{\infty} A_1\left[\nu\left(\varrho^2 -^2 + 2\varrho\varrho^*\right) - \alpha^2\right] d\alpha. \tag{42}$$

By solving the equations, the stresses in the ice sheet can be calculated. Taking the yield stress as the criterion, the ice sheet bending capacity can be computed and it can be determined if the impact force can result in ice breaking by bending failure.

4. Model Validation

The model is validated by comparing the results from a model test performed at Aker Arctic. The details of the ferry model and the ice test condition are described below as well as a summary and discussion of the results in comparison with the computed results from the current study.

4.1. Ice Model Test

The ice model tests for a shuttle passenger ferry were conducted in May 2017 for SSPA Sweden AB at Aker Arctic Technology Inc. [23]. The ferry model is shown in Figure 12. The vessel is designed to operate in the Stockholm archipelago and the main particulars are presented in Table 2. The model was built in scale 1:8.333 and its dimensions and the ice properties can be seen in Table 2. The level ice tests were performed at three speeds according to the specifications given in Table 3, where R_{bc} is ice breaking component and R_{ice} is ice resistance. The total resistance is calculated by dividing the resistance into breaking and submersion components. However, the breaking component in the test is independent of the ship velocity and is defined by the ITTC guidelines [24]. The breaking resistance is the total resistance in level ice minus the resistance in the pre-sawn ice which is assumed to represent the resistance without the breaking component.

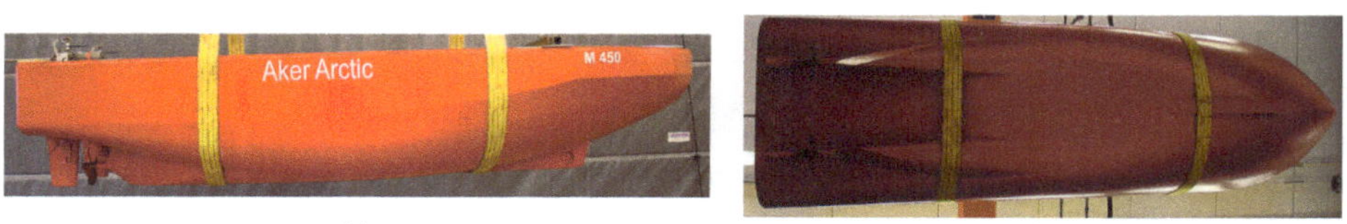

(a) (b)

Figure 12. Side view and bottom view of the ferry model NB550 (In Appendix C of AARC Report A-555): (a) Side view; (b) Bottom view [23]. "Reproduced with permission from [Aker Arctic Technology Inc, SSPA Sweden AB and SLL], published by [Aker Arctic Technology Inc.], 2017".

Table 2. Principal particulars of the ferry NB550 and model ship.

Full-Scale Ship		Model	
L (m)	26.155	Scale factor	8.333
B (m)	7.5	L (m)	3.18
T (m)	2.3	B (m)	0.9
φ_1 (deg)	25	T (m)	0.276
α (deg)	30	μ (1)	0.05
C_b (1)	0.549	E (kPa)	77,979
C_m (1)	0.896	σ_f (kPa)	53.6
Displacement (kg)	247,091	E/σ_f	1456

Table 3. Model test conditions.

No.	V [m/s]	h_i [mm]	σ_f [kPa]	R_{bc} (F_x) [N]	R_{ice} [N]	F_z [N]
1	0.41	26	58.7	16.6	67.2	31.4
2	0.92	26.8	51	44.8	157.6	84.75
3	1.38	30.8	55.1	77.6	240.7	146.8

Photos of the ice breaking patterns are taken during the model tests and are shown in Figure 13. Each color strap on the ruler is 10 cm, which indicates that the broken ice floe field (less than 1 m) is created by the stem and shoulder of the ship model. In Figure 13a, the broken ice floe can be observed and their boundaries can be depicted. It is be assumed that the out layer ice floe fail only once. The stem and shoulder collision can be roughly indicated by the black dot in Figure 13. Figure 13b shows small ice pieces and they may fail more than once. After they first break off due to ice bending failure, they break into more pieces when sliding along the ship hull. The ice breaking length is around 10–30 cm. Multi-contact seems to be happening around the shoulder area. By considering the ship symmetry, the ship geometry, and ice breaking pattern, it is assumed that the stem contact is a single event while the shoulder impact has multiple collision locations ranging from one to four on each side. The simulation is conducted according to such assumption.

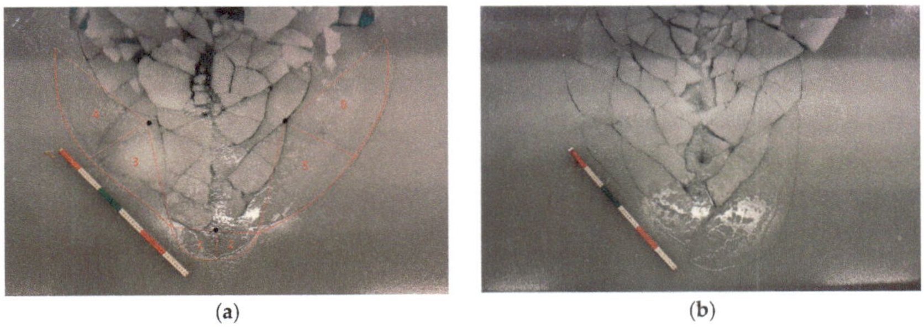

Figure 13. Two examples of Ice breaking pattern captured from ice model test (In Appendix B of AARC Report A-555): (**a**) an estimation of the breaking points indicated by back dots and ice floes marked with numbering for an average ice thickness of 26 mm, (**b**) ice breaking pattern with an average ice thickness of 31mm [23]. "Reproduced with permission from [Aker Arctic Technology Inc, SSPA Sweden AB and SLL], published by [Aker Arctic Technology Inc], 2017".

4.2. Simulation Results

The computation of the impact force on the shuttle ferry is performed by using the ship model data at different speeds. The results are plotted in Figures 14–16 and also given in Table 4. In Figures 14–16, the blue vertical lines indicate the ice force at each time step. The red horizontal lines are the model test data while the black dashed lines are the mean values and the green lines indicate the standard deviation of computed results. It can be seen that the predicted ice force converges to the model test data with the increasing speed. However, the discrepancy is large at the lowest speed, case 1. For the two cases, Cases 2 and 3, a good agreement is observed with an error less than 10%.

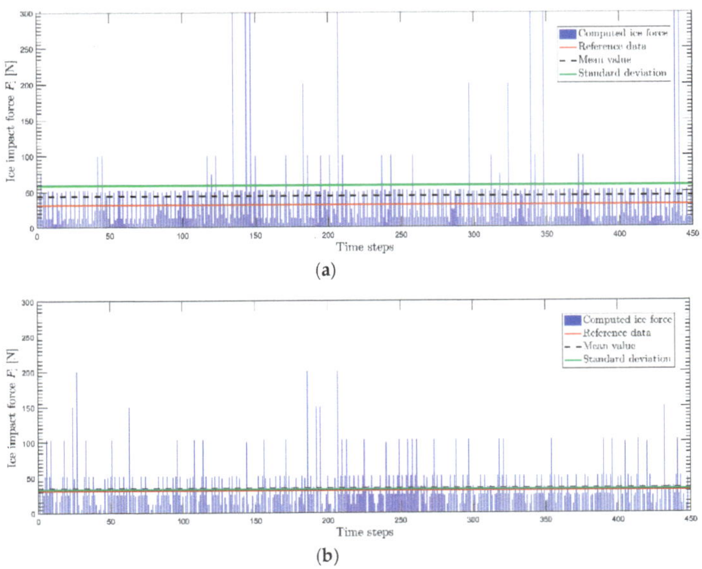

Figure 14. Result comparison for V = 0.41 m/s: (**a**) Case 1: Initial setup; (**b**) Case 1.1: Modified setup.

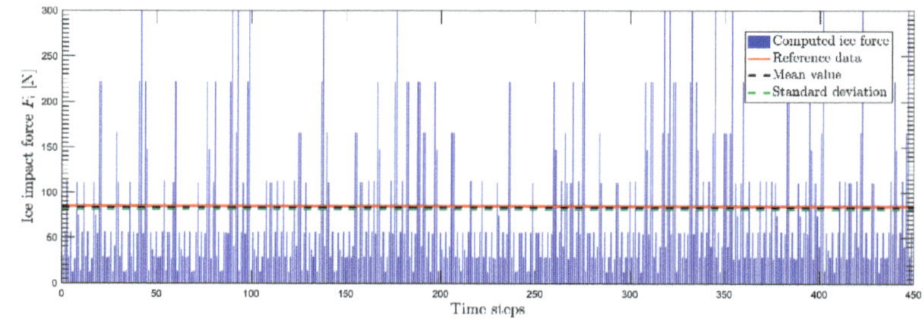

Figure 15. Case 2: Result comparison for speed V = 0.92 m/s.

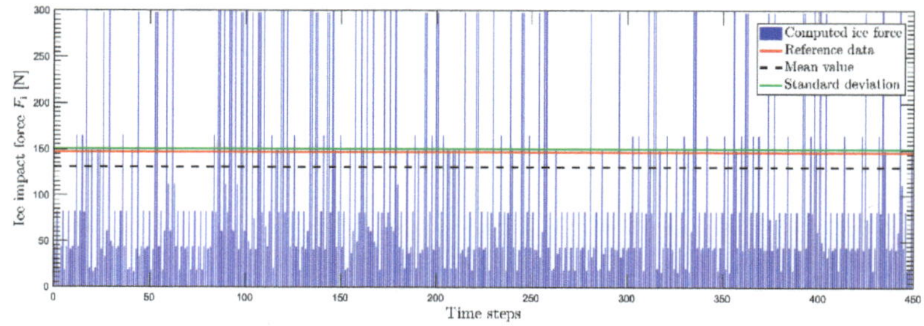

Figure 16. Case 3: Result comparison for speed V = 1.38 m/s.

Table 4. Ice force result summary and comparison.

Case No.	V [m/s]	h_i [mm]	Model Test Data [N]	Average [N]	Error [%]	Standard Deviation [N]	Error [%]
1	0.41	26	31.4	44.1	40.4	58.9	87.58
1.1	0.41	26	31.4	34	8.3	33.6	7.01
2	0.92	26.8	84.75	83.1	1.9	81.3	4.07
3	1.38	30.8	146.8	134	8.7	153.7	4.7

For Case 1, the simulation result does not fit to the test data. A further study is conducted. It is assumed that crushing failure is dominant especially during shoulder collision. It turns out the new simulation, Case 1.1, matches the ice model test data with 8% discrepancy. The modified case demonstrates that the contact scenario, or breaking pattern, affect the outcome considerably. As the breaking pattern and speed relationship is not given in Figure 13, the original assumption may not be applicable for the low speed case. A fraction of the ice impact load history is extracted from Figure 14, as shown in Figure 17. The force pattern matches with Figure 1.

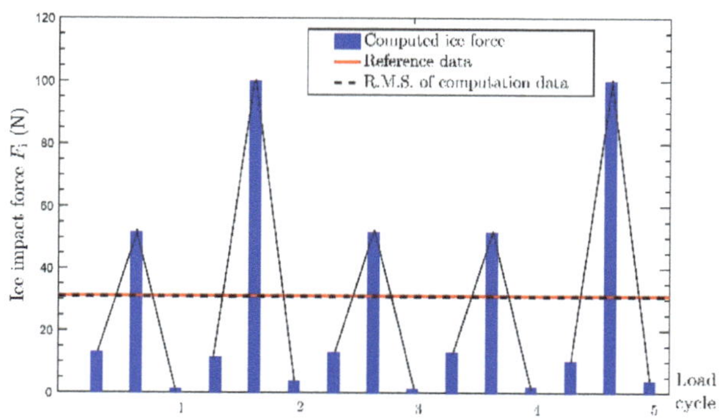

Figure 17. A fraction of ice impact load extracted from Case 1.1.

The ice breaking lengths are calculated and compared with the reference one from empirical Equations (2) and (3), as shown in Table 5. The computed breaking length l_{b1} is larger than the l_{b2}, and both increase with increasing ship speed. Compared to the ice breaking patterns in Figure 13, the computed values of l_{b1} seem reasonable when studying the size of ice pieces in the outside layers from Figure 13. The small ice pieces in the middle are created by the ship body movement when the ice floes slide down along the hull surface, those ice floes deform several thus cannot represent the ice breaking length.

Table 5. Ice breaking length result summary and comparison (unit: m).

	Bow		Shoulder (at B/4)	
No.	l_{b1}	l_{b2}	l_{b1}	l_{b2}
1	0.287	0.113	0.295	0.113
2	0.300	0.116	0.291	0.116
3	0.34	0.128	0.323	0.128

As the different collision locations are evaluated along the ship shoulder, it is interesting to investigate its influence on the discrepancy of the ice forces. Crushing forces in relation to collision locations at different speeds are presented in Figure 18. It clearly seen that the higher speed triggers higher contact forces and the closer to the ship stem, the bigger the force is. Usually, a shoulder collision is only evaluated at location $(x_3, \text{B}/4)$, but Figure 18 can be a good example to illustrate the effect of location on the deciding ice-shoulder impact force.

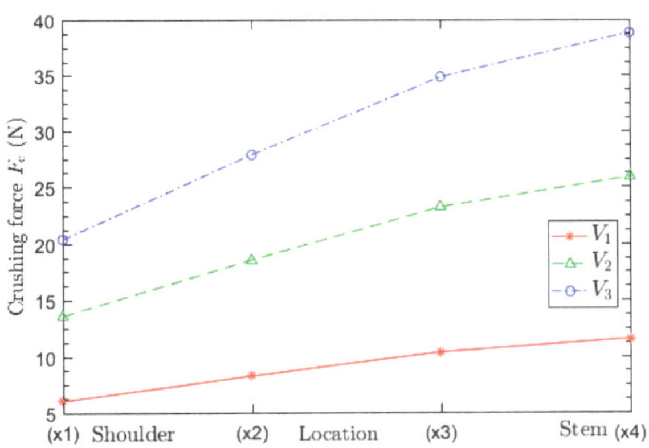

Figure 18. Crushing force in relation to collision location at different speeds.

In order to understand the discrepancy in the predicted force and the model test results at the lower speeds, computed crushing and bending forces are plotted together with measured model test data in Figure 19. The maximum ice bending force and crushing force at $(x_3, \text{B}/4)$ for both the stem and shoulder are selected to make a comparison. The red triangle represents the ice breaking force for the model test data. It can be seen from Figure 19, that the model test data, at the lowest speed, is approximately equivalent to the crushing force, while the second one is in between of crushing and bending force, while the third one, at the highest speed, matches the ice bending force. It can be observed that for low speeds, the ship model test data only indicates crushing not bending. The ice breaking force should be larger with a combination of two components from a statistical point of view. Ship speed also plays an important role in deciding the main failure mode. It can be concluded that the prediction model may be accurate when both crushing and bending occurs and the ice bending force is assumed to be the dominant component in the ice breaking modes.

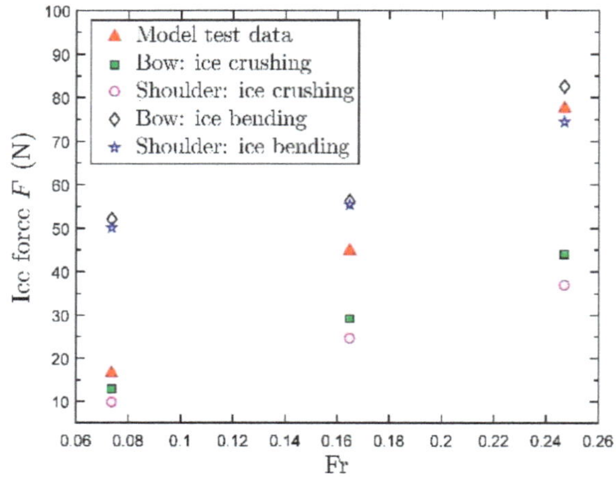

Figure 19. Comparison among model test data and ice failure force components.

In the model test the ice breaking component is calculated in an indirect way using the ITTC guidelines. Considering the result discrepancy, the empirical method should only be used conditionally to compute ice impact force. It is recommended to compared results by ITTC guidelines and simulation, they together can give a good explanation.

5. Case Study

When designing a new ship that is supposed to operate in ice-covered waters it is important to identify the design ice force to assure structural performance. However, it is also crucial to identify the critical ice force as the ship structure needs to survive against extreme loading condition. It can clearly be seen that the critical force is the maximum ice bending breaking force and is significantly larger than the ice impact force. Only the ice critical force, F_C, is considered in this Chapter. The simulation results show that ice bending breaking force is larger than ice crushing force, as seen in Figure 19. Thus, when calculating the ice critical force, the procedure for calculating maximum ice breaking force is used, as shown in Figure 4. A case study is conducted with a full-scale ship to investigate the influence from different design parameters. An IWW barge is chosen as the extreme ice pressure with regards to the ship that was previously evaluated by Zhang et al. [6] using the probabilistic method.

5.1. Barge Information

An IWW barge operating in Lake Mälaren is studied and the ship main particulars are seen in Table 6. The ice properties, e.g., crushing strength and flexural strength, of level ice used in the simulation are given in Table 7.

Table 6. Main particulars of the barge.

L [m]	B [m]	T [m]	Displacement [t]	φ_1 [deg]	φ_2 [deg]	α [deg]	μ
135	11.45	3.4	3938	73	55	45	0.2

Table 7. Ice Properties.

Density [kg/m^3]	E [Pa]	σ_f [kPa]	σ_c [kPa]	h_i [m]
900	10^9	500	2000	0.32

5.2. Parameteric Study

Both the ship design parameters and the ice properties are assumed to have significant effect on the impact load. Some more than others and therefore a parametric study is performed investigating the effect of variations of the rake angle, ship speed, and the ice thickness, see Table 8.

Table 8. Parameters investigated in parametric studies.

Rake Angle	Ship Speed	Ice Thickness
15–75 [deg]	1–5 [m/s]	0.1–0.5 [m]

The critical ice force with respect to ship speed and ice thickness are plotted in Figure 20 for both ship head-on and shoulder collision. It can be seen that the critical force varies with ship speed and ice thickness and the influence is significant. For example, for thinner ice thickness, the ice critical force hardly change with speed. The influence from the speed for thin ice, $h = 0.1$ m, can almost be neglected. When the ice thickness increases, the force increase becomes more and more significant. The figure can be used for guidance in ship operation. For example, assume the structure load capacity is 400 kN, the ship speed can be 2 m/s when ice thickness is 0.5 m or be approximately 5 m/s when ice thickness is 0.4 m.

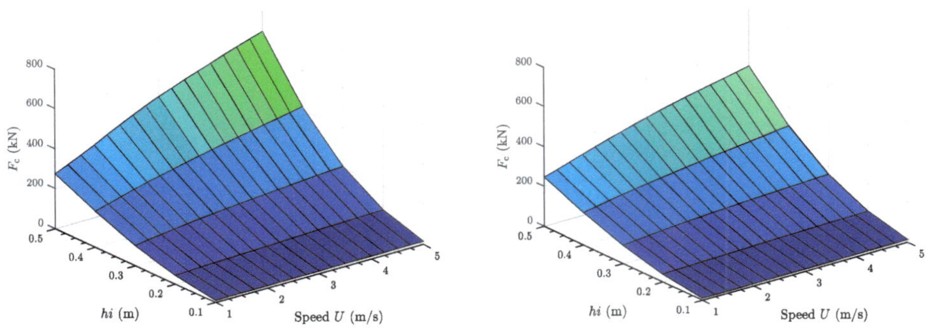

(a) (b)

Figure 20. Ice critical force (F_C) in terms of ship speed and ice thickness for stem angle 25°. (**a**) Ship stem collision; (**b**) Ship shoulder collision.

The parametric study also investigates the influence of the ship stem angle, and the results are plotted in Figure 21. Usually, the stem angle for an ice-going ship is around 20°–25°. The figure shows that when $\varphi = 25°$ the ice critical force takes the lowest value at 1 m/s. However, the critical force increases with ship speed. For a stem angle $\varphi = 75°$ the ice critical force along the whole speed range is low. One explanation can be that the projection to the xy-plane of the contact area and vertical force becomes small based on the ice crushing model. This results in a small ice bending force. The suitable range of the stem angle should be smaller than 15°–75°, for instance, 10°–30°, which causes a larger force to bend the ice sheet. It is suggested that further studies may be conducted as to verify this hypothesis.

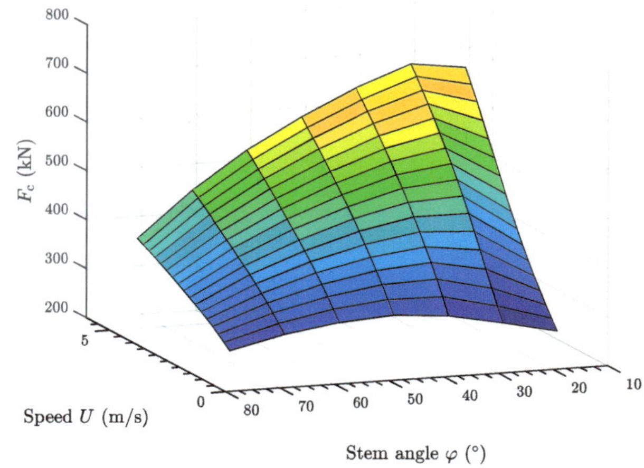

Figure 21. Ship stem collision: ice bending failure force in terms of ship speed and stem angle with ice thickness of 0.32 m.

Ice breaking lengths in relation to ice thickness and ship speed are plotted in Figure 22. It can be clearly seen that the reference breaking length l_{b2} is smaller than the computed one, l_{b1}. However, the difference is decreasing with increasing ice thickness. It can also be deduced that the empirical value l_{b2} is very close to the computed one l_{b1} when ice thickness is approaching 0.5 m and it (l_{b2}) exceed l_{b1} at lower speed. Figure 22 indicates that the Lindqvist [1] method has a limited validity range around ice thickness of half meter.

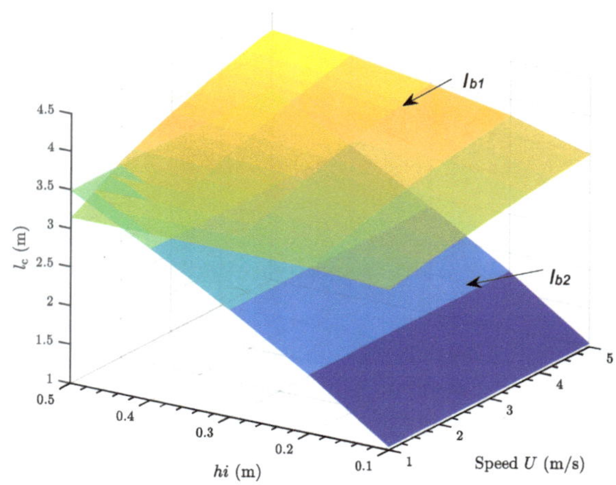

Figure 22. Computed breaking length l_{b1} and reference breaking length l_{b2} for $U = 3$ m/s for bow collision scenario.

6. Discussion and Conclusions

A numerical model for predicting the ice impact force according to ice-hull collision mechanism is developed. The theoretical foundations of the model are explained in detail in this paper. For the realization of such complex ship-ice interaction model several assumptions and simplifications are made. Two ice failure modes, crushing and bending, are thoroughly evaluated. In addition, several different points of collision, i.e., ice ship impact, are considered. The proposed model is simpler than the traditional numerical simulations but still focuses on the essential phenomena and forces, thus, a computationally cost-effective and efficient prediction method is proposed.

6.1. Ice Crushing and Ice Bending

In this work, ice bending is assumed to be dominant over ice crushing. Nevertheless, the assumption does not work every time. Su and Riska [14] found irregularities in resistance values due to changes in the nature of icebreaking pattern: sometimes crushing dominates and sometimes bending. Some parts of the ice sheet are continuously crushed by the vertical parts of the ship hull, e.g., shoulder crushing, without bending failure, then the crushing forces becomes more severe. Croasdale et al. [25] presented a model to calculate the level ice action on wide conical structures. This model is adopted by the ISO Code, ISO/CD19906 [26]. Croasdale's model shows that the contribution of the breaking load component accounts for 20%–30% of the total ice force. Devinder [27] found that ice crushing only occurred at 1% of the time during the observation of Molikpaq ship trial. Still, it caused some of the highest forces. Continuous brittle crushing takes place when an ice sheet moves against a structure at high speed. However, such crushing is mainly happening for relative thin ice. When ice sheet thickness increases, ice bending becomes dominant again. Devinder also found that for low indentation speed and low loading rate, a smooth variation in ice forces based on ice elastic and creep behavior results in ductile deformation [28]. The ejection phase or extrusion (ice sheet drifts away) gives rise to brittle ice failure and the ice force endures an abrupt decrease. Yet, the transition has not been clearly defined as it is ice and scale parameter depended. The energy force used in this model only considers brittle crushing which can cause result discrepancy at low speed. As a rule of thumb, the interaction can be assumed to be bending failure dominant. Another aspect can be the negligence of the different energy components and distributions during the interaction process [29], for instance friction energy.

To identify whether ice crushing or ice bending is dominant, ice crushing force can be plotted through 0–ζ_n together with the bending force. If crushing force is larger at a certain indentation displacement, then bending will not occur.

In numerical simulation, ice is usually regarded as an isotropic material. Still, ice is a non-isotropic, non-homogeneous and orthotropic material. The reasons behind irregularity and discrepancy between model test and simulation should always be kept in mind.

6.2. Ice Impact Load Prediction Model

The ice impact loads for the barge in the present case study was previously studied with the probabilistic method [6], with the predicted design loads 130 kN or 550 kN for ice thickness of 0.32 m depending on different reference dataset. From Figure 20, it can be seen that ice critical force varies from 100–200 kN at same ice thickness. The model in this paper indicates that the predicted impact load of 130 kPa with a contact area of 0.1485 m^2 from empirical methods can be the more valid one. Nevertheless, the ice breaking force of stem collision is higher compared to the shoulder collision, while the contact area of stem collision is lower. This contradicts the conclusion that a smaller contact area results in a higher force according to the probabilistic method. It would be valuable to conduct further investigates on this. The ice breaking force predicted from the numerical model in this paper shows the same magnitude as in other studies, e.g., [12,30].

For the ice model test, ITTC guidelines are used to calculate the ice breaking force. Ice resistance is simplified as a composition of ship resistance in pre-sawn ice and ice-breaking resistance. Pre-sawn ice is level ice in which the approximate breaking pattern is cut prior to the model test. Thus, ice force from model test does not contain the ice breaking component. The guidance has barrier during application.

In the present study, several parameters are studied for ice bending failure. The ship speed and the ice thickness have a positive correlation. The ship speed is shown to have a great impact on the ice breaking in comparison to the other parameters considered. It is also worth mentioning that this parameter is sometimes ignored in empirical equations. The proposed model could assist in quantifying possible differences and improve those equations. For a certain stem angle, the ice loads increase with increasing ship speed. However, the influence from variations in stem angle is not perfectly clear.

The numerical model is sensitive to parameters such as number of wedges, number of collision event. As speed affects ice breaking length, contact case may change. Moreover, when a ship runs slowly, other ice forces, such as sliding or splitting, may become significant. Those possibilities can violate the assumption that bending is dominant and cause discrepancy. Moreover, it is also possible that continuous brittle crushing is happening at such speed. In general, the method works well once suitable parameters are selected. Thus, it facilitates the computation of ice forces for ship design purpose.

The current model indicates that the Lindqvist [1] method is limited regarding the prediction of the ice breaking length. Lindqvist does not work very well for ice thickness below 0.5 m. For IWW light ice conditions, it could be better to find another method.

6.3. Summary

The numerical model introduced herein can be applied in many scenarios. In this paper, the focus is to compute ice impact load based on ice crushing and bending failures. As ice breaking usually occurs at ship bow area, it is taken into consideration by selecting several discretized representative points along the hull. The numerical simulation results are comparable to model test data, which demonstrates that the present mathematical model is rational. The ice impact force calculated with this method can be used to analyze ship structural response, especially in the bow area. The critical force is needed when accessing ship performance under extreme condition. Both global force and local force are calculated from this method. The numerical model can also be adopted to a more global application. More elements can be added to in order to describe the whole waterplane. More geometry contact cases can be analyzed as well for other ice-ship collision circumstances.

A parameter sensitivity analysis of this numerical model is valuable and worth considering in the future work. Moreover, the current model uses computation steps rather than real time steps. Further work can be done to update the time step and generate real time history ice force. Such work would be valuable to study the dynamic structural behaviors. In addition, the proposed model framework can be further developed by including additional ship and ice parameters.

Author Contributions: Conceptualization, M.Z., K.G., and M.B.; methodology, M.Z.; validation, M.Z. and L.Z.; formal analysis, M.Z.; writing—original draft preparation, M.Z.; writing—review and editing, K.G., M.B., and L.Z.; supervision, K.G. and L.Z. All authors have read and agreed to the published version of the manuscript.

Funding: This research was funded by Sjöfartsverket (Swedish Transport Administration): Dnr 16-00778 and Trafikverket (Swedish Transport Administration): TRV 2017/64978. The work is also supported under Joint China-Sweden Mobility Project financed by Swedish Foundation for International Cooperation in Research and Higher Education with grant number Dnr: CH2018-7827 and National Natural Science Foundation of China with grant number 51911530156.

Acknowledgments: We thank SLL, Aker Arctic and SALTECH Consultants AB for their help.

Conflicts of Interest: The authors declare no conflict of interest.

References

1. Lindqvist, G. A straightforward method for calculation of ice resistance of ships. In Proceedings of the 10th Int Port and Ocean Engineering under Arctic Conditions Conference, Luleå, Sweden, 12–16 June 1989.
2. Keinonen, A.J.; Browne, R.P.R.; Reynolds, A. *Ice breaker Characteristics Synthesis*; TP 12812 E; Transportation Development Centre: Calgary, AL, Canada, 1996.
3. Riska, K.; Patey, M.; Kishi, S.; Kamesaki, K. Influence of ice conditions on ship transit times in ice. In Proceedings of the Int Port and Ocean Engineering under Arctic Conditions Conf, Ottawa, ON, Canada, 12–17 August 2001; Volume 2, pp. 729–745.

4. Rahman, M.S.; Taylor, R.S.; Kennedy, A.; Ré, A.S.; Veitch, B. Probabilistic Analysis of Local Ice Loads on a Lifeboat Measured in Full-Scale Field Trials. *J. Offshore Mech. Arct.* **2015**, *137*, 041501. [CrossRef]
5. Taylor, R.S.; Jordaan, I.J.; Li, C.; Sudom, D. Local Design Pressures for Structures in Ice: Analysis of Full-Scale Data. *J. Offshore Mech. Arct.* **2010**, *132*, 031502. [CrossRef]
6. Zhang, M.; Cheemakurthy, H.; Ehlers, S.; von Bock und Polach, F.; Garme, K.; Burman, M. Ice Pressure Prediction Based on the Probabilistic Method for Ice-Going Vessels in Inland Waterways. *J. Offshore Mech. Arct.* **2019**, *41*, 021501. [CrossRef]
7. Daley, C. Energy based ice collision forces. In Proceedings of the 15th International Conference on Port and Ocean Engineering under Arctic Conditions, Helsinki, Finland, 24–29 May 1999.
8. Kerr, A. The Bearing Capacity of Floating Ice Plates Subjected to Static or Quasi-Static Loads. *J. Glaciol.* **1976**, *17*, 76.
9. Lubbad, R.; Løset, S. A numerical model for real-time simulation of ship–ice interaction. *Cold Reg. Sci. Technol.* **2011**, *65*, 111–127. [CrossRef]
10. Valanto, P. The resistance of ships in level ice. *SNAME Trans.* **2001**, *109*, 53–83.
11. Enkvist, E.; Varsta, P.; Riska, K. The Ship-Ice Interaction. In Proceedings of the International Conference on Port and Ocean Engineering under Arctic Conditions, Trondheim, Norway, 13–18 August 1979.
12. Lu, W.; Lubbad, R.; Løset, S. In-plane fracture of an ice floe: A theoretical study on the splitting failure mode. *Cold Reg. Sci. Technol.* **2015**, *110*, 77–101. [CrossRef]
13. Kerr, A.D. *The Bearing Capacity of Floating Ice Plates Subjected to Static or Quasistatic Loads, a Critical Survey*; Research Report; Cold Regions Research and Engineering Laboratory: Hanover, NH, USA, 1975; Volume 333.
14. Su, B.; Riska, K.; Moan, T. A numerical method for the prediction of ship performance in level ice. *Cold Reg. Sci. Technol.* **2010**, *60*, 177–188. [CrossRef]
15. Popov, Y.; Faddeyev, O.; Kheisin, D.; Yalovlev, A. *Strength of Ships Sailing in Ice*; Sudostroenie Publishing House: Leningrad, The Netherlands, 1967.
16. Trafi. Finnish-Swedish Ice Class Rules 2010. In *Ice Class Regulations 2010*; Report No. TRAFI 31298; Finnish Transport Safety Agency: Espoo, Finland, 2010.
17. Daley, C.; Liu, J. *Assessment of Ship Ice Loads in Pack Ice*; ICETECH: Anchorage, AK, USA, 2010.
18. Croasdale, K.R. Ice Forces on fixed, rigid structures. In *CRREL Special Report 80–26, Working Group on Ice Forces on Structures. A State-of-the-art Report*; U.S. Army: Hanover, Germany, 1980.
19. Kerr, A.D.; Palmer, W.T. The Deformations and Stresses in Floating Ice Plates. *Acta Mech.* **1972**, *15*, 57–72. [CrossRef]
20. Kerr, A.D.; Kwak, S.S. The semi-infinite plate on a Winkler base, free along the edge, and subjected to a vertical force. *Arch. Appl. Mech.* **1993**, *63*, 210–218.
21. Nevel, D.E. The Theory of a Narrow Infinite Wedge on an Elastic Foundation. *Trans. Eng. Inst. Can.* **1958**, *2*, 132–140.
22. Nevel, D.E. The narrow free infinite wedge on elastic foundation. *CRREL Res. Rep.* **1961**, *79*, 24.
23. Aker Arctic Technology Inc. *Ice Model Test with a Passenger Ferry for SSPA Sweden*; AARC Report A-555; Aker Arctic Technology Inc.: Helsinki, Finland, 16 June 2017.
24. ITTC. *General Guidance and Introduction to Ice Model Testing*; Technical Report; International Towing Tank Committee: Wuxi, China, September 2017.
25. Croasdale, K.R.; Cammaert, A.B.; Metge, M. A method for the calculation of sheet ice loads on sloping structures. IAHR 94. In Proceedings of the 12th International Symposium on Ice, Trondheim, Norway, 23–26 August 1994; The Norwegian Institute of Technology: Trondheim, Norwegian, 1994; Volume 2, pp. 874–885.
26. ISO/CD 19906. *Petroleum and Natural Gas Industries—Arctic Offshore Structures*: ISOTC 67/SC 7/WG 8 International Standard; International Standardization Organization: Geneva, Switzerland, 2010.
27. Sodhi, D.S.; Takeuchi, T.; Nakazawa, N.; Akagawa, S.; Saeki, H. Medium-scale indentation tests on sea ice at various speeds. *Cold Reg. Sci. Technol.* **1998**, *28*, 161–182. [CrossRef]
28. Sodhi, D.S. Crushing failure during ice–structure interaction. *Eng. Fract. Mech.* **2001**, *68*, 1889–1921. [CrossRef]

29. Blanchet, D.; Kivisild, H.R.; Grinstead, J. Equations for local ice energy dissipations during ship ramming. *Cold Reg. Sci. Technol.* **1990**, *18*, 101–115. [CrossRef]
30. Liu, J.; Lau, M.; Williams, F.M. Mathematical Modeling of Ice-Hull Interaction for Ship Maneuvering in Ice Simulations. *ICE Tech.* **2006**, *845*, 1–8.

© 2020 by the authors. Licensee MDPI, Basel, Switzerland. This article is an open access article distributed under the terms and conditions of the Creative Commons Attribution (CC BY) license (http://creativecommons.org/licenses/by/4.0/).

applied sciences

MDPI

Article

Analysis of a Collision-Energy-Based Method for the Prediction of Ice Loading on Ships

Sabina Idrissova *, Martin Bergström, Spyros E. Hirdaris and Pentti Kujala

School of Engineering, Aalto University, P.O. Box 15300, FI-00076 AALTO Espoo, Finland; martin.bergstrom@aalto.fi (M.B.); spyridon.cheirdaris@aalto.fi (S.E.H.); pentti.kujala@aalto.fi (P.K.)

* Correspondence: sabina.idrissova@aalto.fi; Tel.: +358-41-717-4504

Received: 16 September 2019; Accepted: 22 October 2019; Published: 26 October 2019

check for updates

Abstract: Ships designed for operation in Polar waters must be approved in accordance with the International Code for Ships Operating in Polar Waters (Polar Code), adopted by the International Maritime Organization (IMO). To account for ice loading on ships, the Polar Code includes references to the International Association of Classification Societies' (IACS) Polar Class (PC) standards. For the determination of design ice loads, the PC standards rely upon a method applying the principle of the conservation of momentum and energy in collisions. The method, which is known as the Popov Method, is fundamentally analytical, but because the ship–ice interaction process is complex and not fully understood, its practical applications, including the PC standards, rely upon multiple assumptions. In this study, to help naval architects make better-informed decisions in the design of Arctic ships, and to support progress towards goal-based design, we analyse the effect of the assumptions behind the Popov Method by comparing ice load predictions, calculated by the Method with corresponding full-scale ice load measurements. Our findings indicate that assumptions concerning the modelling of the ship–ice collision scenario, the ship–ice contact geometry and the ice conditions, among others, significantly affect how well the ice load prediction agrees with the measurements.

Keywords: Polar Code; Polar Class; goal-based ship design; arctic ships; energy method; ice load; ice strength

---

1. Introduction

Maritime activity in the Arctic is on the increase, driven by the extraction of Arctic natural resources, trans-Arctic shipping and Arctic tourism. To manage related risks, the maritime industry must ensure that Arctic ships are safe and sustainable. To this end, ships operating in Polar water must be approved under the International Code for Ships Operating in Polar Waters, or the Polar Code [1], enforced by the International Maritime Organization (IMO). The specific objective of the Polar Code is to ensure the same level of safety for ships, persons and the environment in Polar waters, as in other waters [2]. Towards this end, it supplements the International Convention for the Safety of Life at Sea (SOLAS) and the International Convention for the Prevention of Pollution from Ships (MARPOL) to account for Arctic-specific safety hazards, such as sea ice, icing, low temperatures, darkness, high latitude, remoteness, the lack of relevant crew experience and difficult weather conditions.

A ship approved in accordance with the Polar Code is issued a Polar Ship Certificate that classifies the ship as one of the following:

- Category A, for ships allowed to operate in at least medium-thick first-year ice.
- Category B, for ships allowed to operate in at least thin first-year ice.

- Category C, for ships allowed to operate in ice conditions less severe than those included in Categories A or B.

In addition to the ship's category, the ice certificate determines detailed operational limits concerning, for instance, the minimum temperature and the worst ice conditions in which a ship can operate.

The Polar Code is fundamentally goal-based, determining mandatory provisions in terms of goals, functional requirements FR(s), and regulations to meet those goals. As a result, a ship can be approved either as a prescriptive design or as an equivalent design. A prescriptive design is a design that meets all of the prescriptive regulations associated with the FR(s), whereas an equivalent design is a design that is approved in accordance with Regulation 4 of SOLAS Chapter XIV. The latter case results in a so-called alternative or equivalent design. Per regulation 4—"Alternative design and arrangement" of SOLAS Chapter XIV [3], where alternative or equivalent designs or arrangements are proposed, they are to be justified by the following IMO Guidelines:

- "Guidelines for the approval of alternatives and equivalents as provided for in various IMO instruments", MSC.1/Circ.1455 [4].
- "Guidelines on alternative design and arrangements for SOLAS chapters II-1 and III, MSC.1/Circ.1212 [5].
- "Guidelines on alternative design and arrangements for fire safety", MSC/Circ.1002 [6].

A general principle is that any alternative design should be at least as safe as a design determined by prescriptive rules. Many of the prescriptive regulations of the Polar Code, in particular, those related to a vessel's structural strength, include references to the International Association of Classification Societies' (IACS) Polar Class (PC) ice class standards [7]. These consist of in total seven PC notations, ranging from PC 1 (highest) to PC 7 (lowest), corresponding to various levels of operational capability in ice and hull strength. For instance, as per the Polar Code, for a Category A ship to meet FR(s) regarding structural strength, the ship must be constructed in accordance with PC 1–5, whereas a Category B ship must be constructed in accordance with PC 6–7. Alternatively, the scantlings must be determined in accordance with a standard which offers an equivalent level of safety following the above-described principle of design equivalency.

Because the ship–ice interaction process is complex, stochastic and not fully understood, predicting the level of ice loading that a ship will be exposed to is challenging. Multiple different methods to predict ice loading have been proposed, including analytical, numerical and empirical methods [8]. However, to date, none of the existing methods can be considered complete and sufficiently validated to enable them to support direct goal-based structural design following the Polar Code. Thus, in practice, designers are dependent upon the application of the PC standards to obtain regulatory approval for ships designed for operation in ice-infested waters.

For the assessment of design ice loading, the PC standards apply a method based on the principle of the conservation of momentum and energy in collisions, presented by Popov, et al. [9]. The method, in the following referred to as the Popov Method, is relevant for the assessment of ice loading on ships interacting with all types of sea ice. Despite the method being fundamentally analytical, its practical applications, including its application in the PC standards, rely on several assumptions. These concern the modelling of the collision scenario (e.g., type of collision, ship–ice contact geometry), as well as the material properties of ice [10,11]. Because many of the assumptions are empirically determined, the PC standards can be considered semi-empirical, meaning that they might not be efficient when applied on design and operating conditions different from those for which the assumptions were determined [12].

Against this background, this study aims to help designers to make better-informed decisions in the design of Arctic ships by analysing the role and effects of the collision scenario assumptions behind the application of the Popov Method [9]. Specifically, the study addresses the following research questions:

1. How do assumptions behind the modelling of the collision scenario and the description of the operating conditions affect the ice load estimate?
2. What collision scenario assumptions should be applied to obtain reliable ice load estimates for typical operation in level and broken ice?

Because the Popov Method in principle makes it possible to assess ice loading for a wide range of ship designs and operating scenarios, the method is highly relevant in the context of goal-based design. Thus, by addressing the above research questions, we also aim to support progress towards the goal-based design of Arctic ships, as per the above-described principle of equivalent design.

The study is limited to issues concerning ice material properties and non-accidental ship–ice interactions resulting in no or minor structural deformation with relatively thick ice, where the dominant failure mode is crushing. We do not consider the following:

- Issues concerning hydrodynamic effects. These are assumed to have a limited effect on the types of collision scenarios considered [13].
- Issues concerning structural resistance.
- Local plastic deformations. Thus, although the method can do so, we do not consider local plastic deformations, which could affect the obtained load estimate [14,15]. It can be mentioned that allowable plastic deformations are not clearly defined in the PC rules [16].
- Interactions with thin ice, which in addition to crushing would require the consideration of flexural failure and dynamic effects [17]. Daley & Kendrick [18] indicate that when dynamic effects are considered, the design normal force is increasing with increasing impact velocity [6].
- Secondary impacts (i.e., reflected collisions). Daley & Liu [19], among others, demonstrate that when operating in thick ice, reflected collisions may result in critical loads, in particular on the mid-ship area.
- Moving loads along the hull that may result from a collision with pack ice, glacial ice, or ice channel edges are not considered. Kendrick et al. [20] and Quinton & Daley [21] indicate that moving ice loads may cause more damage than stationary loads of similar magnitude. Consideration of moving loads would require the consideration of additional factors such as sliding contact, nonlinear geometric and material behaviours in structural components, requiring the application of the finite element analysis [22].

2. Background

2.1. The Popov Method

The Popov Method models the ship–ice interaction process as an equivalent one-dimensional collision with all motions taking place along the normal to the shell at the point of impact [23]. The model does not consider sliding friction and buoyancy forces. A single hydrodynamic effect, namely the added mass of the surrounding water, is considered.

To simplify the ship–ice interaction process to an equivalent one-dimensional collision, Popov et al. [9] introduce the so-called "reduced mass" concept [23]. Accordingly, the total reduced mass $M_{red}, [t]$ of the ship–ice system is derived by coupling 6-degree of freedom (DOF) equations describing the motions of a ship with 3-DOF equations describing the motions of an ice floe [9]. By assuming that ice floes are of an ellipsoid shape, the 6-DOF and 3-DOF models are converted into a model with a single DOF. As a result, the contact force $F, [MN]$ can be determined as an integral of contact pressure over the nominal contact area $A, \left[m^2\right]$ per

$$M_{red}\frac{\mathrm{d}^2}{\mathrm{d}t^2}\zeta_n(t)\cdot 10^{-3} = F = \int p\left(\zeta_n, \frac{\mathrm{d}}{\mathrm{d}t}\zeta_n(t)\right)dA, [MN] \quad (1)$$

where $\zeta_n, [m]$ is indentation depth, $t, [sec]$ is time, and $p(\zeta_n, \dot{\zeta_n}), [MPa]$ is "contact pressure" [9]. The initial impact conditions at $t = 0$ are as follows: $\zeta_n(0) = 0$ and $\dot{\zeta_n}(0) = v_1$, where v_1 is the reduced speed for the ship $[m/sec]$. At the end of the interaction at $t = t_{max}$ the following conditions apply: $\zeta_n(t_{max}) = \zeta_{max}$ and $\dot{\zeta_n}(0) = 0$.

The ship–ice contact area depends on the hull form, assumed shape of the ice edge and the assumed indentation of the ship into the ice.

$$A = G\zeta_n{}^a, \left[m^2\right] \tag{2}$$

where G is a coefficient depending on the geometric parameters of the ship and the ice, and a is an exponent depending on the assumed shape of the ice edge at the point of impact [9]. This means that the relationship between the normal indentation and the nominal contact area can be determined for any contact geometry.

2.2. Collision Scenario

In an ice floe-structure interaction, there are three limit mechanisms for ice loads: Limit ice strength, limit driving force and limit momentum [24]. The Popov Method is based on the mechanisms of limit momentum, also referred to limit energy mechanism, and it considers impacts between two bodies where one body is initially moving and the other is at rest. The ice load is calculated by equating the available kinetic energy with the energy expended in crushing and the potential energy in collisions.

Due to the non-homogeneous nature of sea ice and the complexity of the ship–ice interaction process, the number of possible ship–ice collision scenarios is large. Thus, when predicting ice loading, assumptions must be made regarding the location of the ship–ice impact (e.g., bow, side, stern), the type of ice involved (e.g., ice floe, ice field), and the resulting contact geometries [11]. Popov et al. [9] consider the following ship–ice collision scenarios:

- An impact between a ship and an ice floe. According to Enkvist, et al. [25] this is the most common collision scenario for merchant ships. As per Table 1, in this scenario, the kinetic energy of the ship $T_1, [J]$ is partially converted into the kinetic energy of the ice floe $T_2, [J]$ and partially consumed in crushing the ice edge $U, [J]$. T_1, T_2, and U are calculated in accordance with the formulas presented in Table 1 (Case 1), where $M_{red}{}^{ship}, [t]$ and $M_{red}{}^{ice}, [t]$ are the reduced masses of the ship and the ice floe, and the parameters $v_1, [m/sec]$ and $v_2, [m/sec]$ correspond to the reduced speeds of the ship and the ice floe.
- An impact between a ship and the edge of an ice field. In this scenario, defined as case two in Table 1, the edge of the ice field is crushed, and the ice field bends due to the vertical component of the contact force $P_v, [N]$. In this scenario, the ship's kinetic energy $T_1, [J]$ is consumed by the crushing $U, [J]$ and by the bending of the ice field $V, [J]$. T_1, V, and U are calculated in accordance with the formulas presented in Table 1 (Case 2), where $f = \frac{P_v}{2 \cdot \sqrt{\gamma_i \cdot D}}[m]$, $\gamma_i, \left[N/m^3\right]$ is the specific weight of ice, $D = \frac{E \cdot h_i^3}{12 \cdot (1-\mu)}, \left[Pa \cdot m^3\right]$ is flexural stiffness of an ice plate, $E, [Pa]$ is the elastic modulus of ice, $h_i, [m]$ is the ice thickness, μ is Poisson ratio for ice.

Table 1. Energy components of the Popov Method as determined based on Popov et al. [9]).

Case 1: Ship–Ice Floe Collision	Case 2: Ship–Ice Field Collision
$T_1 + T_2 = U$	$T_1 = U + V$
$T_1 = 1/2 \cdot M_{red}{}^{ship} \cdot v_1{}^2$	$T_1 = 1/2 \cdot M_{red}{}^{ship} \cdot v_1{}^2$
$T_2 = 1/2 \cdot M_{red}{}^{ice} \cdot v_2{}^2$	$V = 1/2 \cdot f \cdot P_v$
$U = \int_0^{\zeta_n} F d\zeta_n$	$U = \int_0^{\zeta_n} F d\zeta_n$

2.3. Ship–Ice Contact and Ice Crushing Pressure

To manage the complexity of the ship–ice interaction process, ship–ice contact forces can be estimated based on the average pressure and contact area in terms of so-called pressure-area models [26]. These are typically presented in the form of pressure-area relationships $P(A)$ describing the development of the average pressure throughout the ice indentation process [27,28]. Because the ship–ice interaction process is complex and not fully known, $P(A)$ relationships can only be reliably determined based on full-scale measurements.

The maximum force generated by ice acting upon a structure depends on the strength of the ice in the relevant mode of failure [29]. Because in ship–ice interactions, the ice often fails by crushing, ice crushing strength is an important factor for estimating ice loading on ships [30]. As pointed out by Kujala [8], the crushing strength of ice depends on multiple parameters, including ice salinity, ice temperature and loading rate, but all of this is not fully understood. However, as demonstrated empirically by Michel and Blanchet [31], ice crushing strength is proportional to ice compressive strength σ_{co}, $[MPa]$ as per

$$\sigma_{cr} = C \cdot \sigma_{co}, [MN] \tag{3}$$

where C is an indentation coefficient which for the brittle range for sea ice has an empirically determined value of 1.57 [32,33].

2.4. Contact Geometry

To calculate ice loads using the Popov Method, the ship's indentation into the ice must be defined. Following Tunik [11], the indentation can be assessed based on the assumed form of the contact area between the ship and the ice, which depends both on the form of the ship's hull, and on the form of the ice at the point of impact. Per the Popov Method, this form is modelled in terms of an ice–hull overlap geometry, in the following referred to as contact geometry [34].

In real-life ship–ice interactions, many different types of contact geometries occur. Thus, to enable the assessment of ice loading from different types of ship–ice interactions, Popov et al. [9], Daley [34] and Daley & Kim [35] present contact geometry models corresponding to different ship–ice interactions. A selection of these contact geometries are presented in Figure 1, and explained in the following:

- Round contact geometry. This model, which is behind the design scenario of the ice class standards of the Russian Maritime Register of Shipping (RMRS), corresponds to an oblique collision with a rounded ice edge [12]. Related geometrical parameters are the radius of ice floe R, $[m]$, the waterline angle α, $[deg]$ and normal frame angle β', $[deg]$. Parameters α and β' are dependent on the hull form, whereas R is dependent on the ice conditions. In the RMRS ice class rules, R is assumed at 25 m.
- Angular (wedge) contact geometry. This model is applied in the design scenario of the PC rules [14]. Related geometrical parameters are the ice edge opening angle ϕ, $[deg]$, waterline angle α, $[deg]$ and normal frame angle β', $[deg]$. According to Popov, et al. [9], ϕ can be determined by observation of ice segments broken by a ship. Alternatively, ϕ can be determined by numerical simulations [36].
- Symmetric v wedge contact geometry. This model can be applied to assess ice loads from beaching impacts, i.e., when a ship is ramming an ice sheet so that the bow of the ship rises upwards onto the ice sheet [34]. Related geometrical parameters are waterline angle α, $[deg]$, frame angle β, $[deg]$ and stem angle γ, $[deg]$, all of which depend on the hull form at the point of impact.
- Right-apex oblique contact geometry. This model, defined by Daley [34], corresponds to continuous operation in ice without ramming. Related geometrical parameters are waterline angle α, $[deg]$, normal frame angle β', $[deg]$, stem angle γ, $[deg]$ and frame angle β, $[deg]$, all of which depend on the hull form at the point of impact.

- Spherical contact geometry. This model can be applied for the assessment of ice loads on a bulbous bow [20]. Related geometrical parameters are the contact radius r, $[m]$, waterline angle α, $[deg]$ and the normal frame angle β', $[deg]$, all of which depend on the hull form at the point of impact.
- Pyramid contact geometry. This model can be applied to assess ice loads from ship–iceberg interactions resulting in structural deformations [35]. Related geometrical parameters are the ice edge opening angle ϕ, $[deg]$, waterline angle α, $[deg]$ and normal frame angle β', $[deg]$.

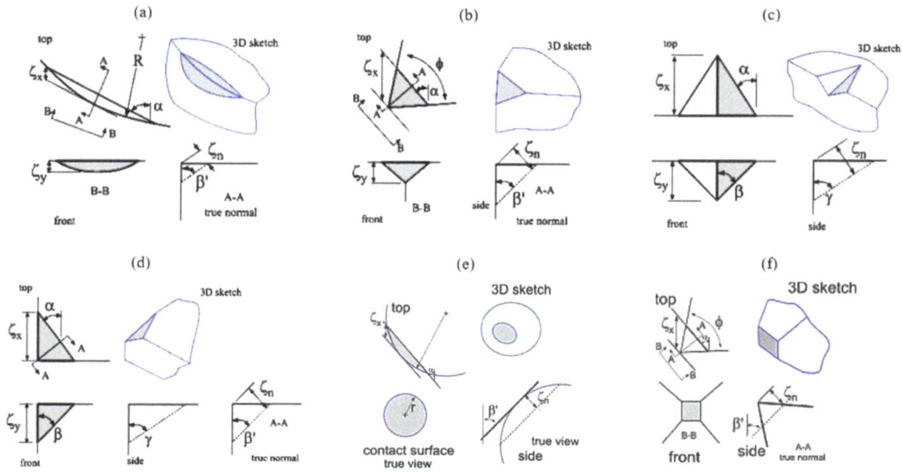

Figure 1. Considered contact geometry models: (**a**) Round; (**b**) Angular; (**c**) Symmetric v wedge; (**d**) Right-apex oblique; (**e**) Spherical; (**f**) Pyramid (Adapted with permission from Daley [34], International Conference on Port and Ocean Engineering under Arctic Conditions, 1999, and from Daley & Kim [35], The American Society of Mechanical Engineers, 2010).

As per the contact geometries presented above, the level of ice loading might depend on the hull form in terms of the frame angle β, waterline angle α and stem angle γ, as shown in Figure 1. Generally, a reduced frame angle, an increased waterline angle, or an increased stem angle result in higher ice loads, and vice versa. However, as discussed by Popov, et al. [9] the influence of the different angles depends upon the point of impact.

2.5. Identification of Assumptions Behind the Popov Method and the PC Rules

Based on Popov et al. [9], Dolny [14], Daley [10], Daley et al. [37] and IACS [38] we identify the following assumptions behind the Popov Method and the related PC rules:

- Assumptions related to the definition of the ship–ice collision scenario. The Popov Method assumes two types of ship–ice collision scenarios defined in Table 1: (a) Ship–ice floe collision, and (b) ship–ice field collision. The PC rules, on the other hand, assume a single type of ship–ice collision scenario, namely a glancing impact with thick level ice.
- Assumptions related to the definition of contact pressure. For typical speeds of ships operating in ice, Popov et al. [9] suggested that the contact pressure $p\left(\zeta_n, \dot{\zeta}_n\right)$ can be assumed equal to the ice crushing strength σ_{cr}, $[MPa]$ over the whole contact area. Accordingly, the maximum ice load corresponding to the maximum indentation depth ζ_{max}, $[m]$ can be defined as per

$$F = \sigma_{cr} \int_0^{\zeta_{max}} dA, [MN] \quad (4)$$

However, this is a simplification, as the crushing strength of ice depends on multiple factors and varies between 1.25 – 10 MPa [9]. In the PC rules, on the other hand, the contact pressure is assumed to correspond to the pressure-area relationship following:

$$P = P_0 A^{ex}, [MPa], \tag{5}$$

where P_0, $[MPa]$ is average pressure acting on 1 m^2 (class dependent), $A, \left[m^2\right]$ is the contact area, and ex is a coefficient empirically determined as −0.1.

- Assumptions related to the definition of the contact geometry case. The application of specific contact geometries requires making assumptions regarding individual contact geometry parameters. When applying the round contact geometry model, Popov et al. [9] recommend the use of R-values in the range of 10–40 m. For angular contact geometry, Popov et al. [9] recommend the use of ice opening angle values (ϕ) in the range of 45°–145°. In the PC rules the assumed contact geometry case is always angular (also referred to as a wedge shape) with a constant ϕ value of 150° [37].
- Assumptions related to the definition of the size of the nominal ship–ice contact area. As per the Popov Method, the nominal ship–ice contact area is assumed to be dependent on the form of the hull and ice at the point of impact as per Equation (2). The PC rules, on the other hand, assuming a wedge-shaped contact geometry, assume a single nominal contact area defined per

$$A = \frac{\zeta_n{}^2 \cdot \tan\left(\frac{\phi}{2}\right)}{(\cos\beta')^2 \cdot \sin\beta'} \left[m^2\right] \tag{6}$$

Naturally, this is a simplification as the actual contact area depends strongly on the assumed contact geometry, and thus on the prevailing ice conditions and the hull form.

- Assumptions related to the definition of load length and load height. The load length and load height, both of which affect the structural requirements in accordance with the PC rules, are derived as a function of the indentation depth [34,39]. In the PC rules, the design load length w is assumed to be defined per

$$w = W_{nom}{}^{wex}, [m] \tag{7}$$

where W_{nom} is the nominal contact length of the assumed wedge-shaped contact geometry and wex is an ice spalling parameter with an empirically determined value of 0.7. The ice spalling parameter accounts for ice edge spalling effects by reducing the size of the contact patch. Although this parameter has a significant effect on the determined scantling requirements, no justification for the assumed value of wex is found in the literature [12]. The design load height h_d is related to the design load length as per

$$h_d = \frac{w}{AR}, [m] \tag{8}$$

where AR is a design patch aspect ratio used to simplify the nominal contact area to an equivalent area of a rectangular patch.

3. Research Method

3.1. Ship–ice Interaction and Material Parameters

We determine the actual ice conditions and loads based on full-scale ice measurements and visual observations of the prevailing ice conditions conducted on the Polar Class 5 classified polar research ship S.A. Agulhas II on a voyage in the Antarctic Ocean in the period November 2013–February 2014 [40]. For the purpose of ice load measurements, the ship's hull was instrumented in accordance with Figure 2 with strain gauges at the bow (frames 134 and 134.5), bow shoulder (frames 112 and 113), and stern shoulder (frames 39.5, 40, 40.5, and 41) [40]. The prevailing ice conditions were measured using a stereo camera.

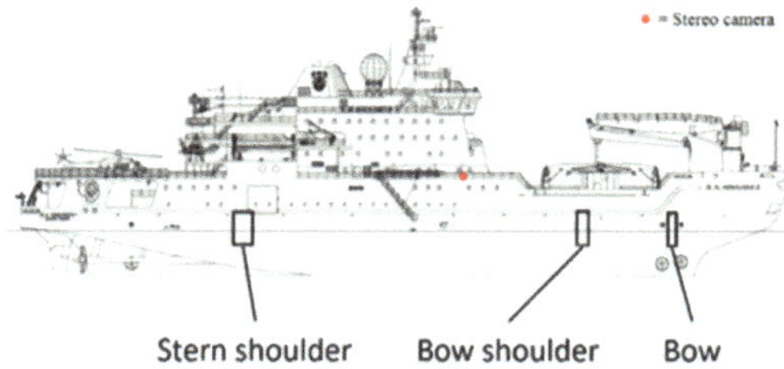

Figure 2. Location of strain gauges on the S.A. Agulhas II. (The figure was determined based on Suominen & Kujala [40]).

The full-scale data describes the ship–ice interaction process and the prevailing operating conditions in terms of [40]:

- Ship parameters including hull particulars: Length, beam, draft, displacement, hull angles.
- Impact parameters including impact location and speed.
- Environmental parameters including ice floe size, ice thickness and air temperature.

From the full-scale data, we extract two data series for two separate operating days representing different types of ice conditions and operating modes:

- Data set A. Measurements from 10.12.2013 representing continuous operation in ice as well as specific manoeuvres such as ramming and reversing in ice. An extract of data set A is presented in Figure 3.
- Data set B. Measurements from 20.12.2013 representing mainly continuous operation in ice. An extract of data set A is presented in Figure 4.

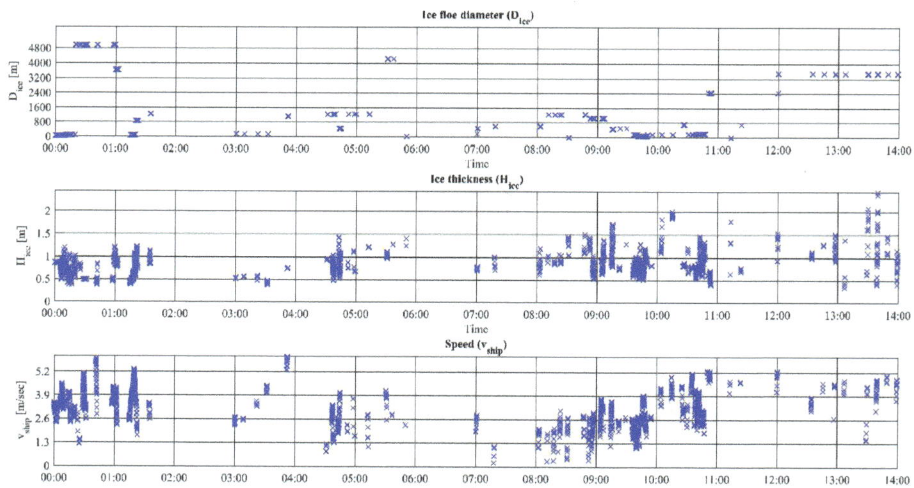

Figure 3. Extract from data set A, showing how the measured ice floe diameter, ice thickness and speed vary during a selected period of 14 h.

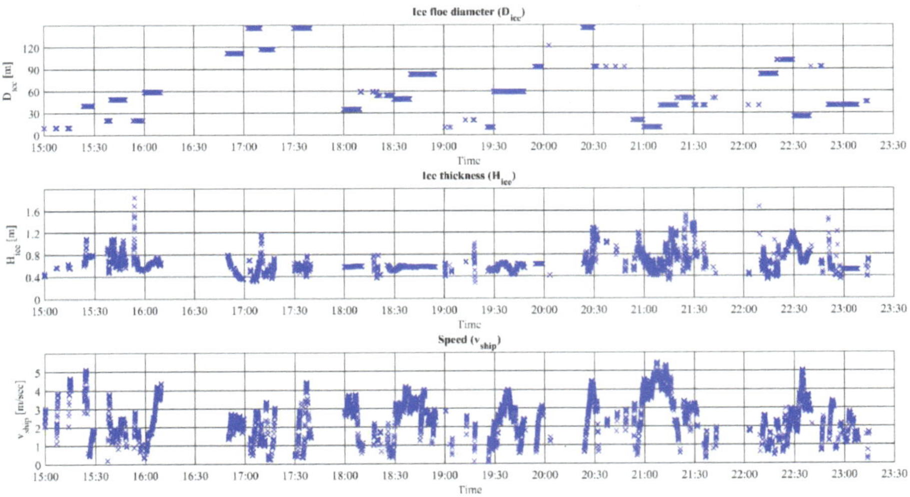

Figure 4. Extract from data set B, showing how the measured ice floe diameter, ice thickness and speed vary during a selected period of 8 h.

The specified ice floe diameters were determined by visual observations from the ship carried out every 10 or 15 min [41]. Because the measurements were carried out by visual observation, and because the size of individual ice floes were specified by categories of diameter range (e.g., 0–20 m, 20–100 m, 100–500 m), the diameter values specified in Figures 3 and 4 are subject to significant uncertainty.

As mentioned in Sec. 2.5, ice crushing strength σ_{cr} is an important factor for estimating ice loading on ships. In order to enable a direct comparison between the measured and calculated ice loads, accounting for the fact that σ_{cr} depends on the prevailing environmental conditions, we calculate σ_{cr} as function of ice salinity, brine volume, ice temperature, ice density and indentation rate in accordance with an approach proposed by Idrissova, et al. [42]. This approach assumes among others that ice crushing strength is proportional to the compressive strength of ice as suggested by Michel & Blanchet [31].

3.2. Collision Scenarios

Following Popov, et al. [9], collision scenarios are defined in terms of the impact location, type of collision (e.g., collision with an ice floe or an ice sheet) and contact geometry. Naturally, these are interlinked, because both the type of collision and the impact location affect the resulting contact geometry. Towards addressing our research questions, we analyse multiple collision scenarios defined by Table 2.

Table 2. Collision scenarios proposed by Popov et al. [9] and Daley [34].

Impact Location	Bow Area, Where the Frame Angle is Larger than 10 [*deg*]
Types of collisions	(a) Glancing impact with an ice floe, (b) Glancing impact with an ice field
Contact geometries	Round, angular, symmetric v wedge, right-apex oblique

Per Table 2, we focus on the bow area because it is typically subjected to the largest ice loads. Specifically, based on the available full-scale data, we analyse ice loading on two specific impact locations in the bow area, namely frames 134 and 134.5. We focus on two of the most common types of collisions, namely glancing impact with an ice floe and glancing impact with an ice field. We consider four different types of contact geometry, namely round, angular (with ice edge opening angle of 145°), symmetric v wedge and right-apex oblique.

We do not consider pyramid and spherical contact geometries, because we assume that these are preliminary intended for the analysis of ship-iceberg and bulbous bow-ice impacts [20,35].

We confirmed this assumption by calculations indicating that the applications of the pyramid and spherical contact geometries result in ice load predictions that are multiple times higher than the corresponding measurements.

3.3. Ice Load Parameters

To calculate ice-crushing pressure p, $[MPA]$, we first need to calculate related ice load parameters including indentation depth, contact area, normal force and load length. We calculate indentation depth directly by the Popov Method for all considered collision scenarios. To this end, we apply the contact geometry specific values for G and a (see Equation (2)) determined by Popov, et al. [9], Daley [34] and Daley & Kim [35]. We then calculate the ship–ice contact area $A, \left[m^2\right]$, following Equation (2), then the related load length l, $[m]$ and the total ice load F, $[MN]$, following Equation (4). The load length is thereby specific for each contact geometry and can be derived as a function of the indentation depth. For example, for round contact geometry, load length l_{round}, $[m]$ is calculated following Equation (9) [9].

$$l_{round} = \frac{5}{3} \cdot \sqrt{\frac{2 \cdot R \cdot \zeta_n}{cos\beta'}}, [m] \tag{9}$$

Based on the calculated parameters, we calculate ice crushing pressure p as a function of the contact area for each considered collision scenario, following

$$p = \frac{F}{l \cdot h}, [MPa] \tag{10}$$

3.4. Line Loads

Equation (1) provides a direct estimate for the total ice impact force, acting normal to the hull shell at the point of impact. The full-scale measurements, on the other hand, correspond to local loads, making it unreasonable to compare them directly [43]. Therefore, to compare the calculated and the measured loads, we first have to convert both into line loads. To this end, we first convert ice loads calculated by the Popov Method into line loads following

$$q = \frac{F}{l} \cdot 10^3, [kN/m] \tag{11}$$

where q, $[kN/m]$ is the line load, F, $[MN]$ is the total ice load on the frame, and l, $[m]$ is the load length that is assumed to correspond to the horizontal length of the contact area.

To convert the full-scale measurements into line loads, we first run the measurements for the selected frames and their combination through a Rayleigh separator, applying a separator value of 0.5 and the threshold value of 10 kN. We then convert the measurements into line loads, separately for the one-frame spacing of 0.4 m (frame 134) and two-frame spacing of 0.8 m (frame 134 + 134.5), as described by Suominen & Kujala [40].

4. Results

4.1. Analysis of Data Set A

Based on the description of the prevailing operating conditions of data set A, we calculate line ice loads for each of the collision scenarios defined in Table 2. An excerpt of the results is presented in Figure 5, comparing measured and calculated line ice loads for two different contact geometry assumptions, namely round and angular. In Figure 6, we compare loads calculated for each collision scenario with the corresponding measurements in terms of 10-min mean, standard deviation (SD) and maximum.

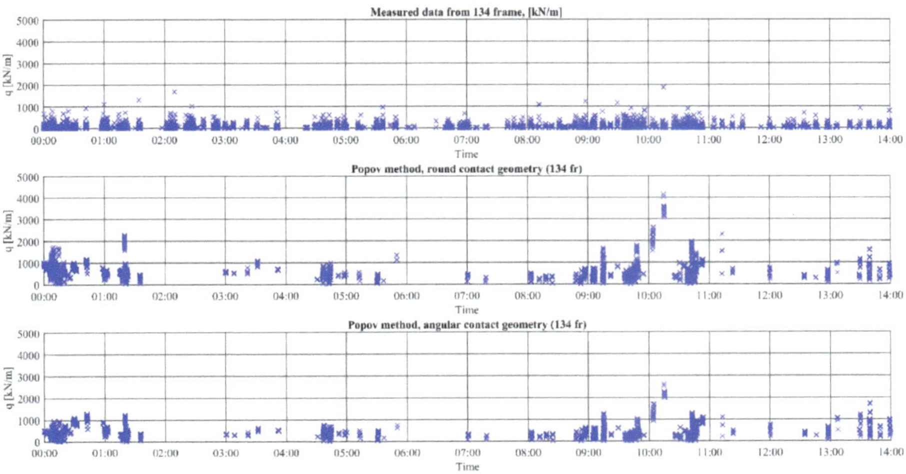

Figure 5. Excerpt of the measured and calculated line ice loads for data set A.

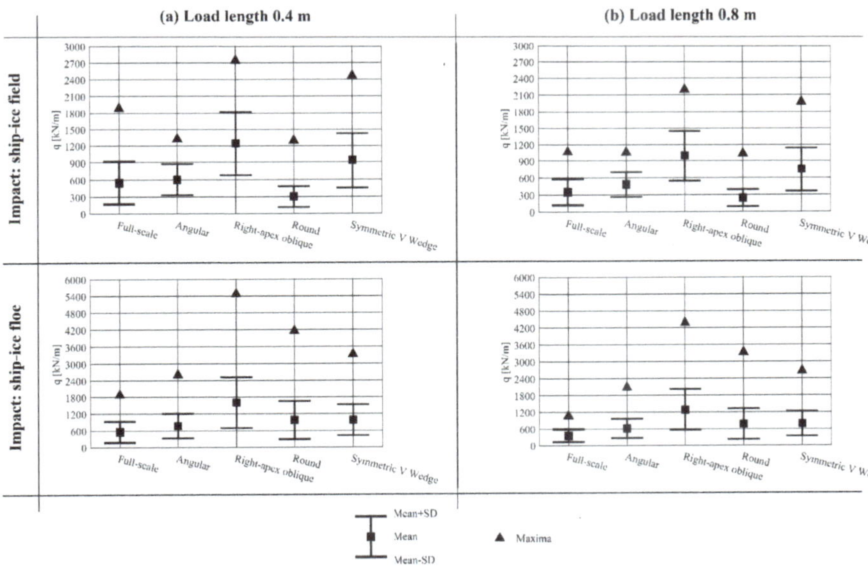

Figure 6. Comparison between loads calculated for different collision scenario assumptions with corresponding full-scale measurements in terms of 10-min mean, standard deviation (SD) and maximum. Load length 0.4 m corresponds to loads on one frame spacing (frame 134), whereas load length 0.8 m corresponds to loads on two frame spacing (frame 134 + 134.5).

Table 3 presents the percentage-deviations between the measured and the calculated 10-min mean Δ_{mean} and extreme Δ_{max} values. As per Table 3, we find that a relatively good agreement is obtained by assuming angular contact geometry in combination with either ship–ice floe impact having a load length of 0.4 m, or a ship–ice field impact with a load length of either 0.4 m or 0.8 m. All other combinations of assumptions result in Δ_{max} and or Δ_{mean} exceeding ± 40%, indicating that their descriptions of the ship–ice interaction significantly deviate from the actual ship–ice interactions behind data set A.

Table 3. Percentage-deviations between loads calculated for various collision scenario assumptions and the measured values of data set A. (Bold font indicates that the deviation is below ±40%.)

Contact Geometry	Ship–Ice Floe Impact				Ship–Ice Field Impact			
	Load Length of 0.4 m		Load Length of 0.8 m		Load Length of 0.4 m		Load Length of 0.8 m	
	Δ_{mean}	Δ_{max}	Δ_{mean}	Δ_{max}	Δ_{mean}	Δ_{max}	Δ_{mean}	Δ_{max}
Angular	**33.10%**	**32.20%**	54.40%	64.60%	**9.50%**	**−34.70%**	**31.80%**	**−0.50%**
Right-apex oblique	98.30%	97.60%	114.40%	121.70%	78.00%	**37.30%**	96.30%	69.30%
Round	55.50%	75.30%	75.60%	102.80%	−58.00%	**−36.80%**	**−36.80%**	**−2.70%**
Symmetric V Wedge	56.20%	76.20%	55.30%	85.40%	52.70%	**27.00%**	73%	59.70%

4.2. Analysis of Data Set B

Based on the description of the prevailing operating conditions of data set B, we calculate corresponding line ice loads for each collision scenario defined in Table 2. An excerpt of the results is presented in Figure 7, comparing measured and calculated line ice loads for four different contact geometry assumptions, namely round and angular, right-apex oblique and symmetric v-wedge. In Figure 8, we compare loads calculated for each collision scenario with the corresponding measurements in terms of 10-min mean, standard deviation (SD) and maximum.

Table 4 presents the percentage-deviations between the measured and the calculated 10-min mean Δ_{mean} and extreme Δ_{max} values. As per Table 4, we find that a relatively good agreement is obtained by assuming either angular or symmetric v-wedge contact geometry in combination with ship–ice floe impact with a load length of 0.4 m. All other combinations of assumptions result in Δ_{max} and or Δ_{mean} exceeding ± 40%, indicating that their descriptions of the ship–ice interaction significantly deviate from the actual ship–ice interactions behind data set B.

Table 4. Percentage-deviations between loads calculated for various collision scenario assumptions and the measured values of data set B. (Bold font indicates that the deviation is below ±40%.)

Contact	Ship–Ice Floe Impact				Ship–Ice Field Impact			
Geometry	Load Length of 0.4 m		Load Length of 0.8 m		Load Length of 0.4 m		Load Length of 0.8 m	
	Δ_{mean}	Δ_{max}	Δ_{mean}	Δ_{max}	Δ_{mean}	Δ_{max}	Δ_{mean}	Δ_{max}
Angular	**−0.80%**	**−36.00%**	**24.20%**	−46.60%	**−14.00%**	−84.90%	**11.00%**	−93.60%
Right-apex oblique	70.30%	**37.30%**	91.30%	**26.60%**	56.30%	**−18.90%**	78.50%	**−29.70%**
Round	54.00%	**13.70%**	76.50%	**2.80%**	**−25.90%**	−86.80%	**−0.90%**	−95.50%
Symmetric V Wedge	**23.40%**	**−12.10%**	47.70%	**−23.00%**	49.70%	**−19.00%**	72.50%	**−29.80%**

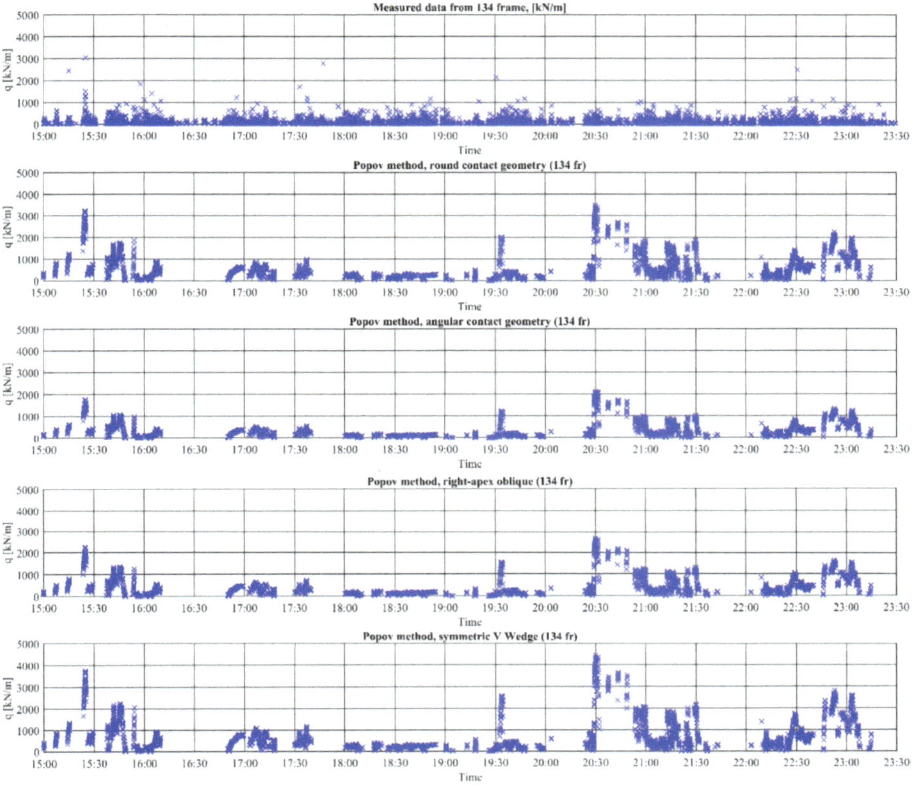

Figure 7. Measured and estimated line loads from ship–ice floe impact for data set B.

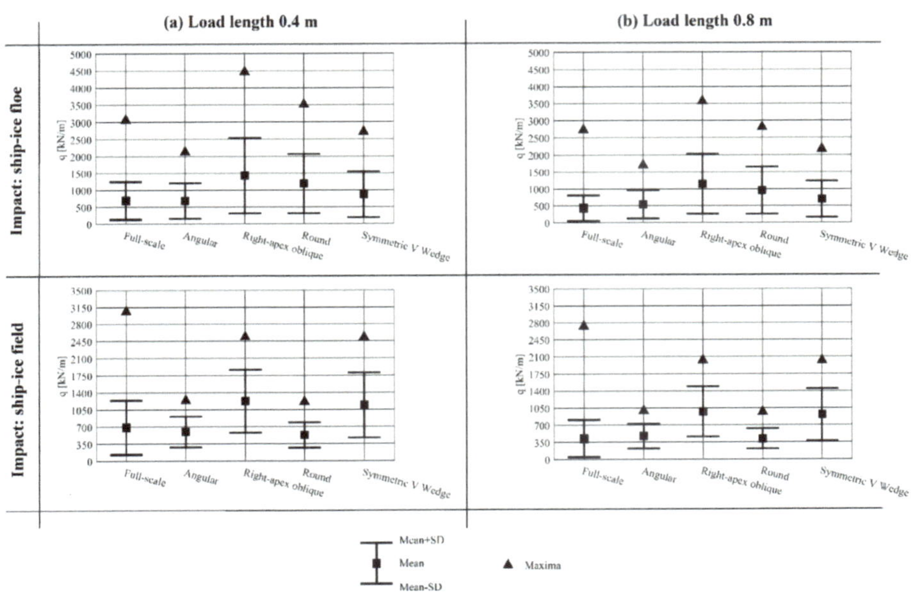

Figure 8. Comparison of measured line loads with estimated by Popov Method for data set B.

4.3. Ice Crushing Strength and Pressure

The considered full-scale ice load measurements (data set A–B) do not include any data to which we can compare our calculated values of ice crushing strength σ_{cr}. Thus, for the validation of our calculated σ_{cr} values, we apply full-scale ice crushing pressure measurements conducted on MV Arctic and IB Kigoriak in Canadian waters, and on IB Sampo in the Baltic Sea [8]. We compare the calculated data points of ice crushing pressure (see Equation (10)) with the full-scale measurements as shown in Figures 9 and 10.

Accordingly, we find that for different collision scenario assumptions, calculated ice crushing pressures p values decrease with the size of the contact area. For both ship–ice field and ship–ice floe impacts, the best correlation with the measurements from the Baltic Sea (IB Sampo) and the Canadian waters (MV Arctic and IB Kigoriak) are obtained by applying angular- and round-contact geometries, respectively. We also find that the calculated pressure points tend to have a higher spread when calculated for an assumed ship–ice floe impact (Figure 10) than when calculated for an assumed ship–ice field impact (Figure 9). This is explained by the fact that in the case of ship–ice floe impacts, the ice load estimate depends on the ice floe size, which varies significantly as shown in Figures 3 and 4.

We calculated the pressure-area points presented in Figures 9 and 10 based on data set B. A similar analysis for data set A resulted in similar findings. It should be pointed out that Popov, et al. [9] do not specify any value of ice crushing pressure. The pressure-area relationship in accordance with PC5 is plotted in Figures 9d and 10d. For comparison, in Figures 9d and 10d we also plot the pressure-area relationship as determined in accordance with PC5. Because the PC rules only consider angular contact geometry, we do not plot any PC definition of ice crushing pressure for other contact geometries. As per Figures 9d and 10d, we find that the slope of the PC5 pressure-area relationship is less steep than that of the pressure-area relationships determined based on full-scale measurements. Specifically, as indicated by the figures, the ice crushing pressures assumed by PC5 appear to be conservative for larger contact areas above approximately 5 m^2.

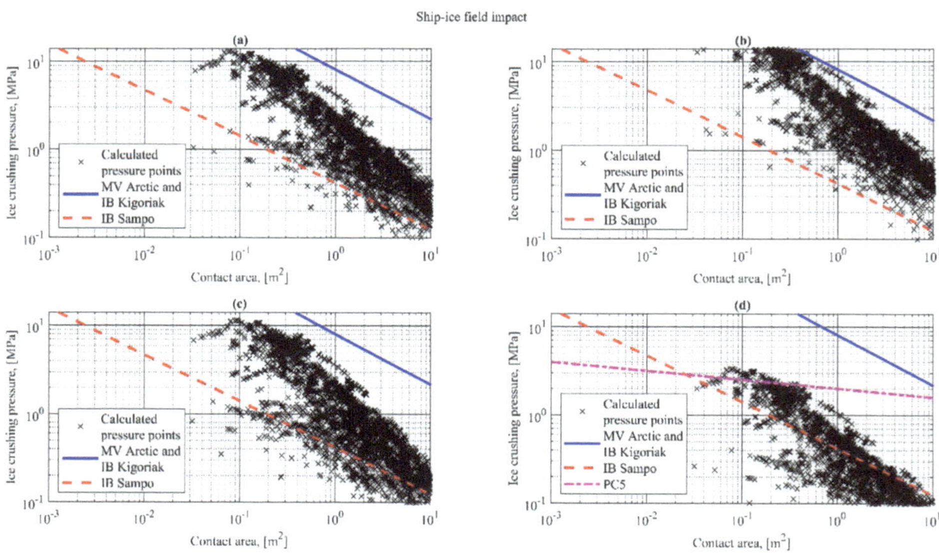

Figure 9. Measured vs. calculated ice crushing pressure-area relationships. Collison type: Ship–ice field impact. Contact geometries: (**a**) Right-apex oblique; (**b**) Round; (**c**) Symmetric v wedge; (**d**) Angular. The ice crushing pressure-area relationships determined based on measurements conducted on MV Arctic and IB Kigoriak, and on IB Sampo, were determined based on Kujala [8].

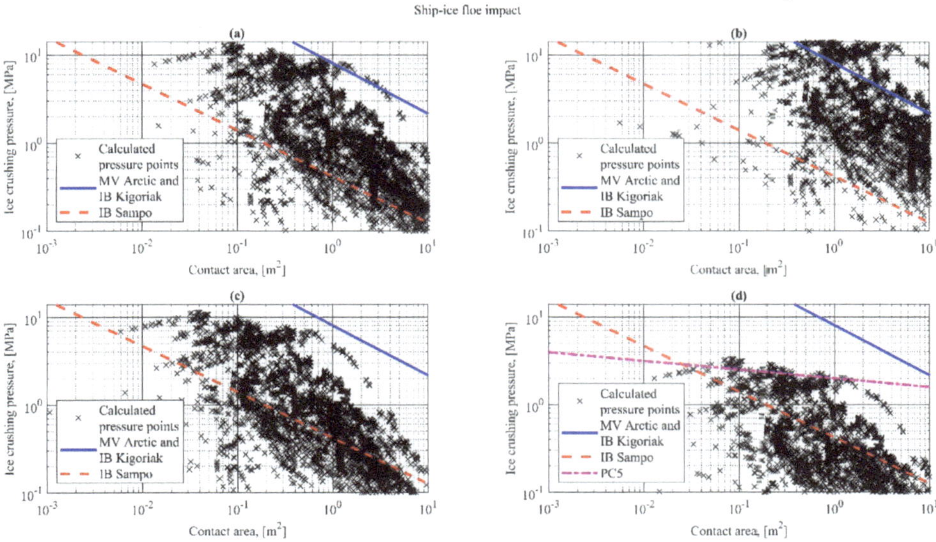

Figure 10. Measured vs. calculated ice crushing pressure-area relationships. Collison type: Ship–ice floe impact. Contact geometries: (**a**) Right-apex oblique; (**b**) Round; (**c**) Symmetric v wedge; (**d**) Angular. The ice crushing pressure-area relationships determined based on measurements conducted on MV Arctic and IB Kigoriak, and on IB Sampo, were determined based on Kujala [8].

4.4. *Influence of the Assumed Ice Edge Opening Angle*

The PC design loads are calculated assuming an angular contact geometry with an ice edge opening angle (ϕ) of 150°. To assess the influence of the assumed ϕ-value, we calculated ice load estimates for ϕ-values between 90°–150°. For the two different load lengths of 0.4 m and 0.8 m, Figure 11 shows how the calculated ice load (*q*), in terms of its mean, standard deviation and maximum, depends on the value of ϕ, and how the calculated values compare to the measurements. We find that calculated ice load decreases with increasing ice edge opening angle (ϕ) values. In addition, by comparing the calculated line loads with the measurements we find that:

- For one-frame spacing (load length = 0.4 m), a good agreement with the mean and maximum of the measurements is obtained for Φ-values around 150° and 110°, respectively.
- For two-frame spacing (load length = 0.8 m), a good agreement with the mean and maximum of the measurements is obtained for Φ-values around 150° and 140°, respectively.

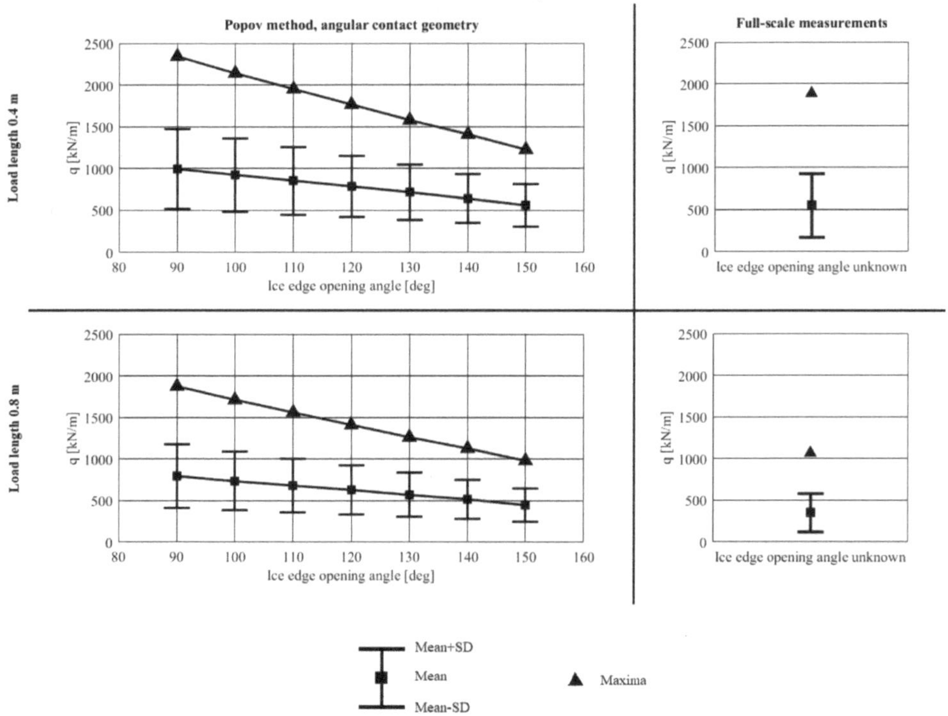

Figure 11. Comparison of line loads from the Popov Method for different ice edge opening angles of angular contact geometry.

5. Discussion and Conclusions

This study analyses a collision-energy-based method for the prediction of ice loading on ships known as the Popov Method. Even though the method is fundamentally analytical, due to unknowns related to the ship–ice interaction process and the material properties of sea ice, its practical application relies on numerous assumptions, some of which are empirically determined. To analyse the role of the assumptions, we compare ice load calculated for various assumptions with two different series of full-scale ice load measurements, referred to as data set A-B. Both of the data series originate from the same voyage, but differ in terms of operating conditions and modes. Data set A includes measurement data from operation in level and broken ice, including specific manoeuvres such as repeated ice ramming and reversing in ice. Data set B, on the other hand, mainly consists of measurement data from operation in broken and level ice without stops.

Towards addressing the research questions of Ch. 1, we calculate ice loads for two different types of collisions, namely the ship–ice floe and ship–ice field collisions. For each assumed collision type, we calculate loads for two different load length assumptions, namely 0.4 m and 0.8 m, and for four types of contact geometries, namely angular, right-apex oblique, round and symmetric v wedge.

In response to RQ1, we find that the obtained ice load prediction is very sensitive to assumptions concerning the modelling of the collision scenario and the operating conditions. For instance, depending on the assumed type of contact geometry, the obtained maximum load estimate might vary between 4%–122% of the corresponding full-scale measurement. This finding indicates that to get a reliable load estimate, the applied assumptions concerning the collision scenario and operating conditions (e.g., type of ice, floe size and ice thickness) must closely correspond to the actual condition.

In practice, this means that when applying the Popov Method, for instance in the context of goal-based design, multiple different ship–ice interaction scenarios must be considered to account for the often complex operating conditions of an Arctic ship.

In response to RQ2, we find that the best agreement with the full-scale measurements is obtained by applying the following combination of the assumptions: (a) Ship–ice floe collision, (b) angular contact geometry, and (c) load length of 0.4 m. Application of these assumptions overestimated the 10-min mean and extreme values of data set A by 33% and 32%, respectively. For data set B, on the other hand, the same assumptions underestimated the 10-min mean and extreme values by 1% and 36%, respectively. All other combinations of assumptions resulted in load estimates where either the mean or the extreme value deviated from either or both of the data series by more than 40%.

When applying an angular contact geometry, it is necessary to assume the value of the ice edge opening angle (ϕ). To investigate the role of this value, we calculated ice load estimates for ϕ-values between 90°–150°. We find that the calculated mean and extreme loads are negatively dependent on the ice edge opening angle. We conclude that the resulting underestimation of loads of data set B using the above defined preferred combination of assumptions could have been avoided by assuming an ice edge opening angle of around 110° instead of the applied value of 145°.

Assumptions behind the Popov Method and the related PC rules do not only concern the ship–ice interaction scenario, but also the material properties of ice, specifically the ice crushing strength. In this study, in order to enable a direct comparison between the calculated and measured loads, we estimate the ice crushing strength directly as a function of the ice salinity, brine volume, ice temperature, ice density and indentation rate. To validate the calculated ice crushing pressure values, we compare them with corresponding full-scale measurements. We find a good agreement between our calculated values and the corresponding measurements, both showing a negative relationship between ice crushing pressure and contact area.

As such, the present study does not aim to influence any existing regulations directly. However, by analysing the effects of the generalizing assumptions behind the PC rules, this work provides motivation and support for a goal-based design in line with the Polar Code. We believe that the present study contributes to a better understanding of the application of the Popov Method in design cases outside of the conditions formally considered in standard PC rules. For instance, to help to assess ice loading in pack ice in ship–ice floe interactions, or to assess ice loads on the aft or mid-body when specific contact geometries must be applied.

Because many of the actual ship–ice interaction parameters related to the measured ice loads, such as contact geometry and load length, are not fully known, the study cannot be considered a validation of the Popov Method. Nevertheless, it is promising that, for specific combinations of the assumptions, we obtained a good agreement with the measurements. Thus, we conclude that the Popov method has the potential to become a useful tool for goal-based design in the context of the Polar Code. It should be noted that this study is limited to assessing the effects of various assumptions when applying the Popov Method to estimate ice loading on ships. As such, it does not aim to directly compare PC design loads with full-scale measurements. Specifying a PC design load requires considering ice loads calculated for at least four sub-regions in the bow area along the waterline [38]. Because our applied full-scale data is limited to ice loads measured on two frames on the bow of the ship, it does not support a fair comparison. Therefore, comparing PC design loads with full-scale measurements remains a topic for future research. Another topic for future research is to address numerous unknowns related to the ship–ice interaction and sea-ice material properties, such as the actual ship–ice contact geometry and line load distribution in various types of ship–ice collisions.

Author Contributions: Formal analysis, S.I.; Methodology, S.I.; Supervision, P.K.; Writing—original draft, S.I. and M.B.; Writing—review & editing, M.B., S.E.H. and P.K.

Funding: This project has received funding from the European Union's Horizon 2020 Research and Innovation Programme under Grant Agreement number: 723526.

Acknowledgments: We thank Mikko Suominen and Roman Repin for the helpful discussion.

Conflicts of Interest: The authors declare no conflict of interest.

Abbreviations

Polar Code	International Code for Ships Operating in Polar Waters
IMO	International Maritime Organization
IACS	International Association of Classification Societies
PC	Polar Class
SOLAS	International Convention for the Safety of Life at Sea
MARPOL	International Convention for the Prevention of Pollution from Ships
FR	Functional Requirements
RMRS	Russian Maritime Register of Shipping
DOF	Degree of Freedom
MV	Motor Vessel
IB	Icebreaker
RQ	Research Question
Notations	
M_{red}	Total reduced mass of the ship–ice system: $[t]$
F	Total ice load, $[MN]$
A	Nominal contact area, $\left[m^2\right]$
ζ_n	Indentation depth, $[m]$
t	Time, $[sec]$
$p\left(\zeta_n, \dot{\zeta}_n\right)$	Contact pressure, $[MPa]$
a	Exponent of the configuration of the ice edge at the point of impact
G	Coefficient of the geometric parameters of the ship and ice
σ_{co}	Ice compressive strength, $[MPa]$
C	Indentation coefficient
T_1	Kinetic energy of the ship, reduced toward the line of impact, $[J]$
T_2	Kinetic energy of the ice, reduced toward the line of impact, $[J]$
U	Ice crushing related work, $[J]$
V	Potential bending strain energy of a semi-infinite ice plate, $[J]$
${M_{red}}^{ship}, {M_{red}}^{ice}$	Reduced masses for the ship and ice floe, [t]
v_1, v_2	Reduced speeds for the ship and ice floe, $[m/sec]$
P_v	Vertical component of contact force, $[N]$
γ_i	Specific weight of ice, $\left[N/m^3\right]$
D	Flexural stiffness of ice plate, $\left[Pa{\cdot}m^3\right]$
E	Elastic modulus of ice, $[Pa]$
h_i	Ice thickness, $[m]$
μ	Poisson ratio for ice
σ_{cr}	Ice crushing strength, $[MPa]$
ζ_{max}	Maximum indentation depth, $[m]$
R	Radius of an ice floe, $[m]$
Φ	Ice edge opening angle, $[deg]$
ζ_x	Projected horizontal indentation depth, $[m]$
ζ_y	Projected vertical indentation depth, $[m]$
α	Waterline angle, $[deg]$
β	Frame angle, $[deg]$
$\beta\prime$	Normal frame angle, $[deg]$
γ	Stem angle, $[deg]$
r	Contact radius, $[m]$
P	Average pressure, $[MPa]$
P_0	Average pressure on 1 m^2, $[MPa]$

ex	Exponent in pressure-area relationship
w	Design contact length, $[m]$
W_{nom}	Nominal contact length, $[m]$
wex	Empirical ice spalling parameter
h_d	Design load height, $[m]$
AR	Design patch aspect ratio
l	Load length, $[m]$
h	Load height, $[m]$
p	Ice crushing pressure, $[MPa]$
q	Line load, $[kN/m]$
l_{round}	Load length for round contact geometry, $[m]$
Δ_{mean}	Difference between the calculated and measured mean values, [%]
Δ_{max}	Difference between the calculated and measured maximum values, [%]

References

1. IMO. *International Code for Ships Operating in Polar Waters (Polar Code)*; International Maritime Organization: London, UK, 2015.
2. Kvålsvold, J. NSR Transit Shipping—A Risk Based Approach. In *Northern Sea Route: New Opportunities*; Det Norske Veritas: Moscow, Russia, 2012.
3. IMO. *Resolution MSC.386(94). Amendments of the International Convention for the Safety of Life at Sea, 1974, as Amended*; International Maritime Organization: London, UK, 2014.
4. IMO. *Guidelines for the Approval of Alternatives and Equivalents as Provided for in Various IMO Instruments*; International Maritime Organization: London, UK, 2013.
5. IMO. *Guidelines on Alternative Design and Arrangements for SOLAS Chapters II-1 and III. MSC.1/Circ.1212*; International Maritime Organization: London, UK, 2006.
6. IMO. *Guidelines on Alternative Design and Arrangements for Fire Safety*; MSC/Circ.1002; International Maritime Organization: London, UK, 2001.
7. Daley, C. Oblique ice collision loads on ships based on energy methods. *Ocean. Eng. Int.* **2001**, *5*, 67–72.
8. Kujala, P. Ice Loading on Ship Hull. In *Encyclopedia of Maritime and Offshore Engineering*; John Wiley & Sons: Hoboken, NJ, USA, 2017.
9. Popov, Y.; Faddeyev, O.; Kheisin, D.; Yalovlev, A. *Strength of Ships Sailing in Ice*; Sudostroenie Publishing House: Leningrad, The Netherlands, 1967.
10. Daley, C. *Background Notes to Design Ice Loads*; IACS Ad-hoc Group on Polar Class Ships, Transport Canada: Ottawa, ON, Canada; Memorial University: St. John's, NL, Canada, 2000.
11. Tunik, A. *Dynamic Ice Loads on a Ship*; International Association for Hydraulic Research: Hamburg, Germany, 1984.
12. Kim, E.; Amdahl, J. Understanding the effect of the assumptions on shell plate thickness for Arctic ships. Trondheim. In Proceedings of the 23rd International Conference on Port and Ocean Engineering under Arctic Conditions, Trondheim, Norway, 14–18 June 2015; pp. 249–261.
13. Huang, Y.; Qiu, W.; Ralph, F.; Fuglem, M. Improved Added Mass Modeling for Ship-Ice Interactions Based on Numerical Results and Analytical Models. In Proceedings of the Offshore Technology Conference, Houston, TX, USA, 5 May 2016.
14. Dolny, J. *Methodology for Defining Technical Safe Speeds for Light Ice-Strengthened Government Vessels Operating in Ice*; Report No. SR—1475; The Ship Structure Committee: Washington, DC, USA, 2018.
15. Storheim, M. Structural Response in Ship-Platform and Ship-Ice Collisions. Ph.D. Thesis, NTNU, Trondheim, Norway, January 2016.
16. Kujala, P.; Goerlandt, F.; Way, B.; Smith, D.; Yang, M.; Khan, F.; Veitch, B. Review of risk-based design for ice-class ships. *Mar. Struct.* **2019**, *63*, 181–195. [CrossRef]
17. Sazidy, M.; Daley, C.; Colbourne, B.; Wang, J. Effect of Ship Speed on Level Ice Edge Breaking. In Proceedings of the ASME 2014 33rd International Conference on Ocean, Offshore and Arctic Engineering, San Francisco, CA, USA; 2014.
18. Daley R&E. *Ice Impact Capability of DRDC Notional Destroyer*; Defence Research and Development Canada: Halifax, NS, Canada, 2015.
19. Daley, C.; Liu, J. *Assesment of Ship Ice Loads in Pack Ice*; ICETECH: Anchorage, AK, USA, 2010.

20. Kendrick, A.; Quinton, B.; Daley, C. Scenario-Based Assessment of Risks to Ice Class Ships. In Proceedings of the Offshore Technology Conference, Houston, TX, USA, 4–7 May 2009.
21. Quinton, B.; Daley, C. Realistic moving ice loads and ship structural response. In Proceedings of the 22nd International Offshore and Polar Engineering Conference, Rhodes, Greece, 17–22 June 2012.
22. Quinton, B.; Daley, C.; Gagnon, R. Guidelines for the nonlinear finite element analysis of hull response to moving loads on ships and offshore structures. *Ships Offshore Struct.* **2017**, *12*, 109–114. [CrossRef]
23. Liu, Z.; Amdahl, J. On multi-planar impact mechanics in ship collisions. *Mar. Struct.* **2018**, *63*, 364–383. [CrossRef]
24. Croasdale, K. The limiting driving force approach to ice loads. In Proceedings of the Offshore Technology Conference, Houston, TX, USA, 7–9 May 1984.
25. Enkvist, E.; Varsta, P.; Riska, K. *The Ship-Ice Interecation;* The International Conference on Port and Ocean Engineering under Arctic Conditions: Trondheim, Norway, 1979.
26. Daley, K.; Riska, K. *Review of Ship-ice Interaction Mechanics Report from Finnish-Canadian Joint Research Project No. 5 "Ship Interaction With Actual Ice Conditions" Interim Report on Task 1A;* Report M-102; Ship Lab, Helsinki University of Technology: Espoo, Finland, 1990.
27. Frederking, R. The Local Pressure-Area Relation in Ship Impact with Ice. In Proceedings of the 15th International Conference on Port and Ocean Engineering under Arctic Conditions, Helsinki, Finland, 23–27 August 1999.
28. Masterson, D.; Frederking, R. Local contact pressures in ship/ice and structure/ice interactions. *Cold Reg. Sci. Technol.* **1993**, *21*, 169–185. [CrossRef]
29. Croasdale, K.R. Crushing strength of Arctic ice. In Proceedings of the Coast and Shelf of the Beaufort Sea, Symposium, San Francisco, CA, USA, 7–9 January 1974.
30. Riska, K. Ice edge failure process and modelling ice pressure. In *Encyclopedia of Maritime and Offshore Engineering;* John Wiley & Sons: Hoboken, NJ, USA, 2018.
31. Michel, B.; Blanchet, D. Indentation of an S2 floating ice sheet in the brittle range. *Ann. Glaciol.* **1983**, *4*, 180–187. [CrossRef]
32. Michel, B. *Ice Mechanics;* Laval University Press: Quebec, QC, Canada, 1978.
33. Kheysin, D. *Opredeleniye Vneshnikh Nagruzok Deystvuyushchikh na Korpus Sudna pri Ledovom Szhatii [Determination of External Loads Which Act on the Hull of A Ship When under Pressure from Ice];* Problemy Apktiki i Antapktiki: Leningrad, Russian, 1961; p. 7.
34. Daley, C. Energy based ice collision forces. In Proceedings of the 15th International Conference on Port and Ocean Engineering under Arctic Conditions, Helsinki, Finland, 24–29 May 1999.
35. Daley, C.; Kim, H. Ice collision forces considering structural deformation. In Proceedings of the ASME 2010 29th International Conference on Ocean, Offshore and Arctic Engineering, Shanghai, China, 6–11 June 2010; pp. 1–9.
36. Li, F.; Kotilainen, M.; Goerlandt, F.; Kujala, P. An extended ice failure model to improve the fidelity of icebreaking pattern in numerical simulation of ship performance in level ice. *Ocean Eng.* **2019**, *176*, 169–183. [CrossRef]
37. Daley, D.; Dolny, J.; Daley, K. *Safe Speed Assessment of DRDC Notional Destroyer in Ice, Phase 2 of Ice Capability Assessment;* Defence Research and Development Canada: Halifax, NS, Canada, 2017.
38. IACS. *Requirements Concerning POLAR CLASS;* International Association of Classification Societies: London, UK, 2016.
39. Kõrgesaar, M.; Kujala, P. *Validation of the Preliminary Assessment Regarding the Operational Restrictions of Ships Ice-Strengthened in Accordance with the Finnish-Swedish Ice Classes When Sailing in Ice Conditions in Polar Waters;* Aalto University publication series Science + Technology: Espoo, Finland, 2017.
40. Suominen, M.; Kujala, P. The measured line load as a function of the load length in the Antarctic waters. In Proceedings of the 23rd International Conference on Port and Ocean Engineering under Arctic Conditions, Trondheim, Norway, 14–18 June 2015.
41. Suominen, M.; Bekker, A.; Kujala, P.; Soal, K.; Lensu, M. Visual Antarctic Sea Ice Condition Observations during Austral Summers 2012–2016. In Proceedings of the 24 International Conference on Port and Ocean Engineering under Arctic Conditions, Busan, Korea, 11–16 June 2017.

42. Idrissova, S.; Kujala, P.; Repin, R.; Li, F. The study of the Popov method for estimation of ice loads on ship's hull using full-scale data from the Antarctic sea. In Proceedings of the 25th International conference on Port and Ocean Engineering under Arctic Conditions, Delft, The Netherlands, 9–13 June 2019.
43. Suominen, M.; Su, B.; Kujala, P.; Moan, T. Comparison of measured and simulated short term ice loads on ship hull. In Proceedings of the 22nd International Conference on Port and Ocean Engineering under Arctic Conditions, Espoo, Finland, 9–13 June 2013.

© 2019 by the authors. Licensee MDPI, Basel, Switzerland. This article is an open access article distributed under the terms and conditions of the Creative Commons Attribution (CC BY) license (http://creativecommons.org/licenses/by/4.0/).

MDPI
St. Alban-Anlage 66
4052 Basel
Switzerland
Tel. +41 61 683 77 34
Fax +41 61 302 89 18
www.mdpi.com

Applied Sciences Editorial Office
E-mail: applsci@mdpi.com
www.mdpi.com/journal/applsci

www.ingramcontent.com/pod-product-compliance
Lightning Source LLC
Chambersburg PA
CBHW042010260926
51850CB00020B/2962
* 9 7 8 3 0 3 6 5 1 9 1 9 7 *